LOBBYING THE EUROPEAN COMMISSION

For Carl and Michael Kerigan

Lobbying the European Commission

The case of air transport

DINOS KYROU
Lecturer and Research Fellow, The University of Aberdeen
Research Fellow, The Robert Gordon University, Aberdeen

Routledge
Taylor & Francis Group
LONDON AND NEW YORK

First published 2000 by Ashgate Publishing

Reissued 2018 by Routledge
2 Park Square, Milton Park, Abingdon, Oxon, OX14 4RN
52 Vanderbilt Avenue, New York, NY 10017

Routledge is an imprint of the Taylor & Francis Group, an informa business

ISBN 13: 978-1-138-33932-3 (hbk)
ISBN 13: 978-0-429-44112-7 (ebk)

Contents

List of Tables

List of Abbreviations

AA	American Airlines
ACI	Airports' Council International
ADV	Arbeitsgemeinschaft Deutscher Verkehrsflughafen
AEA	Association of European Airlines
AEAC	Association des Constructeurs Européens d'Automobiles
ALFA-ACI	Aeroports de Langue Française Associés á ACI (the Association of French Language Airports of ACI)
AOA	(UK) Airport Operators Association
APOC	Anglo Persian Oil Company
BA	British Airways
BAA	BAA (formerly the British Airports Authority)
BBC	British Broadcasting Corporation
BEA	British European Airways
BM	British Midland Airways
BOAC	British Overseas Airways Corporation
CAP	Common Agricultural Policy
CBI	Confederation of British Industry
C.E.	Communautés Européenes
CMLR	Common Market Law Review
COREPER	Committee of Permanent Representatives of the EU
CRS	Computer Reservation System
DG	Directorate General of the European Commission
DGIV	Directorate General for Competition
DGVII	Directorate General for Transport
EACEM	European Association for Consumer Electronics Manufacturers
EATP	European Association for Textile Polyolefins
EC	European Community
ECAC	European Civil Aviation Conference
ECJ	European Court of Justice
ECR	European Court Reports
EEA	European Economic Area
EEC	European Economic Community

EFPIA	European Federation of Pharmaceutical Industry Associations
ERA	European Regional Airlines' Association
EU	European Union
FAA	Federal Aviation Administration
FFPs	Frequent Flyer Programmes
F/N	Footnote
HACAN	Heathrow Association for the Control of Aircraft Noise
IATA	International Air Transport Association
ICAO	International Civil Aviation Organisation
IGHA	Independent Ground Handlers' Association
JAA	Joint Aviation Authorities
LCD	Lowest Common Denominator
MEP	Member of the European Parliament
O.J.	Official Journal (of the European Communities)
ÖTV	Öffenntlicher Transport Verband (German Public Transportation Union)
QMV	Qualified Majority Voting
SAGB	Senior Advisory Group Biotechnology
SAS	Scandinavian Airlines System
TEU	Treaty on European Union
UCCEGA	Les Aeroports Français (The French Airports)
UNICE	Union of Industrial and Employers Confederations of Europe

Preface & Acknowledgements

Bob Ayling resigned from his position as Chief Executive of British Airways on 10th March 2000. The press argued that there were at least two reasons for his decision; the 1997 cabin crew strikes, and the possible imminent announcement of a financial loss for the airline. Bob Ayling jumped before he was pushed, it was widely claimed. If true, the pushers would have been the institutional share-holders who control a large chunk of BA (as well as many other large firms listed on the London Stock Exchange), and the non-executive directors who were clearly getting nervous about the bad press that both Ayling and BA were getting. However, let us not forget that only a few days later these same institutional share-holders were valuing 'Lastminute.com', an internet firm with the turnover of a petrol station and losses of over 1.5 million Euro (US$1.6m) a month, at over 1 billion Euro. Indeed, it was in the long-term interests of such share-holders and directors that Ayling seemed to have been acting. The recent losses were typical of most airlines; the 1998 collapse of the south-east Asian economies led to massive over capacity and a slashing of economy fares (Virgin Atlantic, a rival to BA, claimed that it had made a profit during this time. However, as Virgin does not publish its accounts, establishing the validity of this claim is somewhat difficult). Further, oil prices had more than doubled in 1999 (a flight from Heathrow to Manchester, only 200 miles, uses over 16 000 Euro of fuel). Ayling decided that the future profitability of BA lay, not in attracting low-yield economy passengers, but in selling to the right kind of passengers, i.e. the high yield, business passengers, or the full fare economy passengers; much better to do this, and to improve profitability, than to go for growth in absolute passenger numbers. Ironically, this strategy is praised by those in the financial centre of London and by those who follow air transport economics; yet it was these very people who were about to give Ayling the push. Indeed, he has set the ground work for the future profitability of his former airline, but has paid for its past under-performance, despite the fact that, other than maybe softening his stance to the all important air transport unions, there is little more he could have done. Little, other than one important matter which could not have been more badly handled. This issue was the failure of BA to form a trans-Atlantic air transport 'alliance' with the US carrier American Airlines.

Originally announced in 1996, but still today only partly developed due to the conditions stipulated by the European Commission (the 'executive' of the European Union), this was one area where Ayling was let down by those who advised him and the then Conservative government. That administration was keen to turn the matter into a 'keep your nose out of it Brussels', political issue. Unfortunately for British Airways, Ayling allowed himself to fall into this trap. Although supremely skilful at lobbying UK national governments (particularly the Labour administration), he manifestly failed to apply this domestic talent to the new EU arena. The key issue surrounding this matter was the allocation of airport landing and take-off slots, a matter integral to this book, which will be fully explained in later chapters. Indeed, the slot allocation issue is one which will have enormous ramifications on BA's future strategy. Only two months after the departure of Ayling, his successor to the post, the Australian Rod Eddington, proposed a full-scale merger with BA's European competitor, KLM of the Netherlands. Such a deal would mean that BA and KLM would have an enormous proportion of the EU air transport market far beyond what was proposed almost four years earlier with American Airlines. It would be unlikely, to put it mildly, that the European Commission would even consider this proposal without, at the very least, demanding the 'surrender' of airport slots, primarily at Heathrow (the busiest airport in the world in terms of international traffic), but also at Amsterdam, at least equal in number to those demanded as a price for the proposed alliance with American Airlines. If I am proved wrong following the publication of this book it will, I am certain, be only as a result of a massive shift in the thinking and strategic direction of the competition section of the European Commission. I will take the gamble to become a hostage to fortune (never a good idea), and say that the probability of the European Commission allowing the creation of a BA / KLM 'super airline' is zero.

Only a few years ago air transport was very simple; it was a stitch-up between different countries. The passengers paid enormous fares, the airlines received enormous subsidies from their governments and ran enormously non-profitable routes for their own national prestige. All this was pointed out by Pillai in his groundbreaking work *The Air Net* in 1969, but it has taken years for anything to be done about this sorry state. Back then, who would have thought that British Midland, until recently a quietly successful UK airline would today be challenging for the right to compete on trans-Atlantic routes? However, still today, on the verge of the centenary of the first powered flight, intergovernmental procrastination, upholding the plainly anti-competitive and archaic 'Bermuda II' agreement, prevents BM from entering this market (Bermuda II allows only two US carriers, at present American Airlines and United Airlines, and two UK carriers, at present BA and Virgin Atlantic, to operate between the US and Heathrow. By a remarkable coincidence, the fares of these four airlines, which are up to 40% more than those from continental hubs such as Paris and Amsterdam, are often identical). Nevertheless, the fact that BM is attempting to fly on such routes at all is an indication that the tight reins on air transport have been loosened (albeit by not very much), by reluctant national governments.

There is a myth, which continues to exist, that the American domestic market, still the largest domestic air transport market in the world, was somehow free from these anti-competitive devices; just try explaining that to someone buying a domestic ticket in the US on anything other than one of the 'low cost' carriers, such as Southwest Airlines. A typical one way fare from New York to Miami would set you back over US$900! The truth is that regulatory authorities the world over, including those within the US, have been happy for an unsatisfactory situation to continue for flights both between different countries and within their own territory.

This book examines two aspects of attempted air transport deregulation within what has become the best hope of liberalisation, the European Union. Yet hurdles remain, and they are not small. The slot allocation issue, mentioned above, was due to be sorted out by January 1997. Yet seven months into the following century the competition and transport sections of the European Commission, seem still to be at loggerheads over the issue. The result has been inertia and confusion, affecting not only the airlines, but also the most neglected individual in all this, the airline passenger. The second case study relates to the attempts by various airports within the EU to hang on to one of their last cash-cows; the monopoly provision of ground-handling services. As narrow as these two cases sound, they are indicative of the resistance to change that has been displayed by national governments the world over.

However, the EU is, arguably, the most liberal market for air transport in the world. This is no thanks to those reluctant Member States who have constantly dragged their feet over the issue, but is instead due to the vigorousness of the European Court of Justice, upholding and clarifying the very competition law that many of these Member States tried so hard to forget that they had ever signed up to.

In order to force reluctant countries to liberalise, or to stitch-up the market even further (depending on what side of the fence you are on), airlines have entered into 'alliances' with one another. These alliances can broadly be described as types of mergers, but without a majority control of one airline by another, when regulation prevents this (as it does for trans-Atlantic tie-ups). However, 'full blown' alliances (those where fares are formally set between the partner airlines), require strict regulatory approval. Such alliances constantly shift in nature, and will have undoubtedly have altered to some degree by the time this book has been printed. Indeed, the alliance between the German 'flag carrier' (or 'traditional' national airline), Lufthansa, and the American carrier United and others, the so called 'Star' alliance, has recently been joined by British Midland, its attraction to Star being its enormous number of take-off and landing slots at Heathrow (Singapore Airlines is now also part of the Star Alliance but Virgin Atlantic, with which it has an alliance, is confusingly, not part of Star. Singapore Airlines purchased a 49% stake in Virgin Atlantic at the end of 1999).

The 'One World' alliance, consisting of British Airways and American Airlines, amongst others, is not yet in full swing. The primary reason for this is that at the centre of One World are BA and AA; these two airlines have not been allowed to fare set for reasons which may have little to do with the potential anti-competitive nature of the proposed agreement and more to do with the appalling way that the deal has been 'sold' to the regulators in Washington DC and in Brussels; both turned up their noses at the proposal and, as a result, it is an alliance which is never likely to reach the perfect cruising altitude. This book could rightly be accused of focusing too much on the EU angle in this case. However, this is as a result of the fact that the BA / AA proposal is integral to the whole EU debate covering slot allocation. Besides, if BA cannot get the EU to agree to its position, it has little hope for agreement in Washington; in a trans-Atlantic negotiation it does help to have your own people (i.e. the European Commission) on your own side!

If I could justly be accused of neglecting an area in this book it is the role that charter airlines play in air transport; their role is manifestly underestimated by both industry observers and legislators alike. The first port of call for EU officials is always the airports and the flag carrier airlines, whilst the views of charter carriers are considered almost as an afterthought. As a result, charter airlines played almost no role in the examined case studies of this book. I blame, amongst others, the EU legislators and the air transport press for this. Furthermore, the 'interest groups' which lobby (or, more politely, 'represent') airlines to EU officials could do a much better job of their representation. Further, the low cost carriers such as 'easyJet', and the specialist carriers such as Virgin Atlantic, do not even have sufficient associations to put forward their views (this is a particularly damning indictment as the 'most successful' of the airline associations – the Association of European Airlines, which almost exclusively represents the flag carriers, receives, as this book reveals, a high level of criticism from both legislators and from one of its key members). This situation is reflective of the fact that the different air transport interest groups need to sort out how their views should be represented to EU legislators and officials in this developing 'post state-controlled' era of air transport. Until this happens all airlines, but particularly the charter, low cost and specialist airlines, will continue to be inadequately represented and the views of the airports, who have got their act together in terms of lobbying ability, will continue to dominate the legislative discussion.

Imperative to this book has been the application of EU, competition law (some legal purists still insist on the term 'EC' law, but it has recently become the norm to refer to EU law). Foremost among the articles of the Treaty of Rome referred to are the provisions on competition. The Amsterdam Treaty, which was finally ratified by the 15 Member States of the EU in 1999, unhelpfully re-numbered these provisions. The numbers I have referred to in the book are the old numbers, as they were the ones in use during the period that the study was undertaken. For those of you who wish to keep up-to-date, the all important Article 86 has become Article 82 and Article 85 has become Article 81. Article 90 has become Article 86, Article 89 has become Article 85 and Article 88 has become Article 84. Despite the gallant attempts of the Member States to bemuse an already somewhat confused populace of the EU by these changes, the actual provisions of the competition articles remain the same, although the old DGIV (the 'Competition Directorate' of the European Commission), has suggested new arrangements to allow the Member States to individually execute some aspects of competition policy. Indeed the current situation of the centralisation of competition policy within Brussels leads to an over-stretched administration of competition law. As a result, it is very often the big 'news worthy' issues, such as airlines battling for airport slots, that become a main concern of the competition section of the European Commission, whilst more 'minor' infringements may go unnoticed.

I must also add that the Directorate Generals of the European Commission are now officially no more, as a result of attempts by the European Commission's new boss, Romano Prodi, to 'simplify' things. Subsequently, DGIV has now been renamed, simply, 'Competition', and the old DGVII, Transport, has been amalgamated with the Energy section and is subsequently named 'Energy and Transport. The Commissioners are now Mario Monti of Italy and Loyola de Palacio of Spain, respectively.

Finally I would like to thank those who helped me with my research and with the preparation of this book. The list is much too long to show here, but I would especially like to thank Pat Kyrou, Andreas Kyrou, and Maria Kerigan for their time and effort. In particular, I am enormously indebted to my PhD supervisor, Professor Trevor Salmon, at the University of Aberdeen, for his invaluable insights and tremendous assistance.

Dinos Kyrou
University of Aberdeen
July 2000

Introduction

Few would argue with the remark that the development of international air transport has actively contributed towards a form of 'interdependence' between states, by facilitating the flow of goods and people across national boundaries (see: Keohane and Nye, 1975, pp.363-409; Haas, 1976, pp.173-212; 1980, pp.357-405; Reynolds and McKinlay, 1979, pp.141-169). However, the subject of international air transport also consists of a combination of security aspects which are generally associated with the traditional 'realist' models of international relations (Jönsson, 1987, p.4). Since World War 1, the first war in which aircraft played any role (initially in reconnaissance and then combat), aviation has been intrinsically linked with the national defence implications of states (Cooper, 1947, pp.49-84). As WW1 was the catalyst that developed aviation, air power, unlike sea power, did not evolve from commercial means of transportation, but evolved and was developed for the purposes of war. As a direct result, national governments became the principal actors in the development of aviation.

In addition to national defence, air transport became intrinsically linked with foreign policy considerations. Commonly, one of the first actions of newly de-colonised states was the establishment of a publicly subsidised, national airline, regardless of any economic justification for such an act. The creation of a 'flag carrier' airline was, and continues to be seen by many states as a crucial element in creating a national identity.

The regime which administers global air transport, the International Air Transport Association (IATA), codified this state centric view of air transport within the annexes of the 1944 'Chicago Convention'.[1] The Convention established IATA as a 'price fixing' administration where states would bilaterally 'fix' air transport tariffs for their respective flag carriers. Governments would arrange all elements of air travel between their respective states on a bilateral basis, and the flag carriers were the only airlines able to operate pre-arranged routings. For example, the political administrations of the UK and Japan would decide at what time their flag carriers would fly between London and Tokyo, the capacity allocated for the nationalities on the aircraft (typically 50% British, 50% Japanese), and the fares to be charged.[2] Subsequently, the world air transport system was (and largely continues to be), a 'cartel', where governments carved up the market and largely escaped the rigors of competition (Pillai, 1969).[3] The European Economic Community was no exception to this global rule and the price fixing situation in its Member States reflected IATA's global practice. Most Member States created a flag carrier airline; for example, Air France and Lufthansa in France and Germany respectively,[4] and the governments of these Member States would fix air tariffs on a bilateral basis. This situation continued for over thirty years after the creation of the European Communities, despite the fact that such practices would appear to violate the competition provisions contained within the Treaty of Rome.[5]

The following chapters deal with specific case studies concerning aspects of integration which have taken place in the air transport sector within the European Community / European Union. This is not to say that there has been no resistance to this integration from various Member States. Moreover, it is no coincidence that the Member States that continue to publicly own and attempt to financially subsidise their flag carriers (such as France and Ireland, as well as the 'olive belt' states, Italy, Greece, Spain and Portugal), are the same Member States that feel that air transport is more closely linked to traditional notions of foreign policy and national identity. It is more often these Member States that are the most resistant to the harmonisation and integration of air transport policy, and the subjection of air transport to the norms of EC competition law. Moreover, those states that see the profitability of 'their airlines' as a primary concern (such as the UK and the Netherlands), were amongst the first to privatise their flag carriers and were normally the keenest to integrate air transport policy within the competition provisions of the Treaty of Rome. In short, some Member States of the EU view air transport as a business, whilst others continue to see the issue as a matter of national identity that must be protected.[6] It is partly because of this latter view that the process of air transport liberalisation did not begin until 1987. Liberalisation was a direct result of a case before the European Court of Justice in 1986,[7] supported by a further case in 1989.[8] These court cases determined that, under EC law, air transport could not be treated in a different or preferential way to other industries. It was therefore ruled that many of IATA's practices violated the competition provisions of the Treaty of Rome. Subsequently, the Member States were obliged to remove these 'barriers to trade' in the air transport sector.[9] This process of air transport liberalisation was introduced in stages and it was not until 1992 that the last of the 'three packages' of liberalisation[10] was finalised as EC legislation. Furthermore, since that time it has become apparent that the removal of such barriers to trade in the air transport industry has been insufficient to facilitate effective competition. There have subsequently been attempts to create frameworks and structures in order to enhance this liberalisation that was completed *de jure* in 1992.

The case studies of this book - an examination of a European Commission Proposal for a Council Directive and the attempt by the Commission to formulate a Proposal for a Council Regulation - are examinations of stages in the process of policy formulation which are aimed at enhancing this liberalisation. In both of the examined case studies the Competition Directorate of the European Commission (DGIV), has sought to substantially expand its own authority and competence in the air transport areas of 'ground-handling' and 'slot-allocation.'[11] Both case studies reflect the increasingly prominent role that DGIV has begun to play in the last decade. This increasing prominence may well be reflective of what Stephen George describes as a restoration of the Commission's 'pride and self confidence' which began upon the appointment of Roy Jenkins as its president in 1976 (George, 1992, pp.14-15). Alternatively, it may be the case that this increasingly prominent role of DGIV is not so much reflective of a 'restoration', but is indeed a 'creation' of pride and self confidence within this unique *quasi-judicial* branch of the European Commission.[12] Indeed, followers of *neo-functionalism* may well argue that the evidence presented in the forthcoming chapters concerning the inflated role of DGIV is indeed evidence of this theory.

Neo-functionalism & spillover

Chapter 1 will examine in detail the theory of *liberal intergovernmentalism*, a form of which (later described as *neo-liberal intergovernmentalism*), is applied to the case studies of air transport contained within this book. The origins of liberal intergovernmentalism can be found in the *neo-functionalist / neo-realist* debates.[13] Unlike neorealism, neo-functionalism argues that the actions of states affect all other states, to a greater or lesser degree, in the international system. Additionally, states, argue neo-functionalists, are not the only actors in the international system. The actions of states are influenced by interest groups in a wider society than that which exists within the domestic political arena; furthermore, such interest groups may also transcend national boundaries (George, 1992, pp.20-21; Keohane and Nye, 1975; Lindberg and Scheingold, 1971, p.105).[14]

Neo-functionalists believe that such activity between interest groups promotes *spillover* where integration in one level of activity can lead to further integration in another level of activity.[15] Such *spillover* can take three main forms:

1. Actors (such as individuals, 'interest groups', or firms), can shift their activities to a new centre for their own self interest; this is defined as *political spillover*.

2. Institutions, such as the European Commission, or the Court of Justice and the Court First Instance, can act as 'midwives' of integration, pushing integration forward; this is defined as *cultivated spillover.*

3. Interdependence (for example between firms in the EU), results in policy makers from each Member State being unable to separately formulate effective policy in a particular area. The resulting necessity for integration is defined as *functional spillover* (Tranholm-Mikkelsen, 1991, pp.1-22).

It may be argued that the liberalisation of air transport within the European Union reflected *cultivated spillover,* as the process was pursued by the Commission and applied by the European Court of Justice, both acting as 'midwives' of integration. Evidence does indeed suggest that the landmark case, which 'forced' Member States to adopt liberalisation, *Nouvelles Frontières* (1986)[16] was an example of such cultivated spillover. It has been suggested that it was DGIV, as well as the Directorate General for Transport, DGVII, that persuaded the French travel agent Nouvelles Frontières to confront the French Ministère Public when the French government attempted to prevent this travel agent from setting airline fares below a national minimum level. DGIV may have done this in order to establish this key legal precedent; such a precedent was viewed by DGIV as being the only way in which air transport competition could be forced upon the reluctant Member States.[17]

Competition issues aside, the gradual integration of the air transport regulatory authorities of the Member States into a unitary, EU wide authority is an example of *Functional spillover*. The increase of air traffic over the EU has resulted in a substantial amount of difficulty in co-ordinating many different nationally based air traffic control systems. A single air traffic control unit for the EU (as well as various other states), known as 'Eurocontrol' has been established to attempt to cope with the increased demands faced by national air traffic controllers. This organisation is an example of integration arising from the functional necessity of a particular task to be conducted at an integrated, rather than at a national level. Another organisation, the Joint Aviation Authorities (JAA), conducts some of the activities which were previously carried out at the Member State level by national safety authorities, such as the UK Civil Aviation Authority. Commentators argue that a single authority is necessary for safety controls to be effectively carried out throughout Europe. However, the JAA does not yet possess the mandate to execute its full range of potential powers, primarily because of political resistance from certain Member States. Even functional spillover - by definition an activity resulting from the technical necessity of integration - can be resisted for political reasons.[18]

Political spillover has clearly occurred in many areas of EC policy making. For example, Chapter 2 details the powers that the Commission has recently acquired in the area of merger policy. Interests that would have ordinarily targeted the national level have little choice but to target the European Commission, primarily the Directorate responsible within the Commission, DGIV. The Chief Executive of Dowty Aerospace, a major manufacturer of aircraft components, recently explained how he was able to persuade the Competition Commissioner of the potential benefits to competition that would arise from a merger between his own firm and a French aerospace company. He was successful despite the fact that his legal advisors had said that his company would be not be. When asked by the author why Dowty Aerospace had been successful, where other notable UK air transport interests have more recently failed, the Chief Executive replied:

> Brussels is something that is there and has to be dealt with. Firms cannot ignore Brussels in such situations. If they fail to target the relevant person in Brussels and concentrate only on their national governments they are not going to succeed. This is why we intensively lobbied the Competition Commissioner, even taking him various models of landing gear to show him.[19]

At the UK national level, such an interest would need to persuade the relevant regulatory authority[20] that such a merger would not prevent competition within the state, and indeed would enhance the UK's competitive abilities at the international level. However, in the above example, it needed to be explained to the Competition Commissioner how the merger would not prevent competition at the European level rather than the national level, and how the merger would enhance the EU's 'competitive edge' in relation to the US, etc. Therefore, not only must the actor shift its activity as a result of political spillover, but the methodology employed by the actor must also change.

The hypotheses and the examined case studies

Case Study No.1 The formulation of a Commission Proposal for a Council Directive on airport ground-handling

This case study examines the formulation of a Commission Proposal concerned with the 'the liberalisation of airport ground-handling'.[21] The study focuses on the actors (and their representative groups), that wished to achieve certain policy goals pertaining to the Proposal. In particular, the study examines the Proposal making dynamics of the Commission that existed throughout the formulation of the Proposal.

Case Study No.2 The intended formulation of a Commission Proposal for a revised Council Regulation on slot-allocation

As stated in a 'temporary' 1993 Council Regulation,[22] a Commission Proposal for a revised Regulation was to be formulated no later than 1 January 1996.[23] This case study focuses on the reasons why the Commission was unable to formulate a Proposal, despite being legally bound to do so by EC legislation.

The case studies are focused around two hypotheses.

> Hypothesis No.1
>
> Liberal intergovernmentalism argues that the position of each European Union Member State on a policy matter is influenced by 'elite interests' within each Member State (this is described as an 'elite accommodation'). The Member State subsequently represents the elite interest in the Council of Ministers.
>
> Hypothesis No.1 of this book proposes that liberal intergovernmentalism may not only be reflected at the Member State level, but that the properties that comprise liberal intergovernmentalism are also reflected within the European Commission; this is subsequently defined as a *neo-liberal intergovernmentalism.* Subsequently, Commission actions are influenced by the elite interests within the Member States.

If the hypothesis proves to be correct, then although political spillover may have occurred (i.e. actors have shifted their representation from the national sphere), the *dynamics of policy formulation* (i.e. the response of targets to the eliteness of interests), may not have changed.

The second hypothesis tests interest group theory. Primarily, it tests the relevance of 'elite interests' within the interest group structure, and the decision-making procedures of groups.

> Hypothesis No.2
>
> For an 'interest group' to be successful in its representation to the Commission it must be highly effective in formulating its position on the relevant issue; 1. In particular, *group cohesion* is a necessary pre-requisite in order to achieve such successful representation. To achieve this cohesion, and in order to avoid a *lowest common denominator* of group decision making, the method of decision making must be on a majority basis, rather than one based on unanimity; 2. Additionally, it is the large firms (which are categorised as elite interests), which must be preferentially accommodated within a group in order to ensure this group cohesion, and ultimately, its successful representation.

The chapters

Chapter 1, *The Theoretical and Historical Considerations of Interest Representation* has three primary objectives.

First, the Chapter sets out to introduce the notion that there exist certain 'elite interests' within states, which, because of their political importance, ultimately influence a state's position on an issue when in negotiation with other states (see: Moravcsik, 1993, pp.473-524; Putnam, 1988, pp.427-461; Hix, 1994, pp.1-30 and Huelshoff, 1994, pp.255-279). This notion is the primary theoretical background to the two case studies of Commission action. Section 1 also examines the supporting theories of 'two level games', 'elite accommodation', and 'reverse state strength' theory.

Section 2 sets out to establish a comprehensive working definition of an elite interest, in order that an accurate comparison in the case studies can be made, and a measurement of their relative success in influencing the actions of the European Commission can be established. To achieve this, Section 2 examines the background to 'national champion' interest representation; the most common property of an elite air transport interest throughout is that such elite interests are, in the main, national champion firms.

Sections 1 and 2 are essentially examinations of elite and non-elite interest representation in order to provide a theoretical background for Hypothesis No.1. Section 3, however, explores aspects of interest group theory to provide the necessary background for the examination of Hypothesis No.2 (a hypothesis based on interest group decision making). In particular, examples of interest group representation, where the representation has arguably been improved by re-structuring the format and decision making of the group will be examined. The theories reviewed in Section 3 acknowledge that interest group representation can be undermined by firms within the group operating 'diverse tactics' which side-step the organisation and thereby reduce its effectiveness (McLaughlin, Jordan, Maloney, 1993, p.211). Section 3 examines the theory that group representation must primarily take into account the views of the largest firms within the group (the elites), if interest representation by the group is to be successful; without this, interest groups are 'doomed to failure', in the phraseology of Greenwood (1994, p.195). Section 3 also examines various abilities possessed by elites to successfully represent their interests to policy formulators. It concludes that elites, primarily because of greater resources in comparison to non-elites, have the ability to represent themselves at three levels. Level 1 is the representation to their Member State, Level II is representation via the interest group, whilst Level III is representation directly to EU institutions (for example, to the European Commission, and European Parliament). Non-elite firms are reliant on Level II interest representation as they do not have the resources (in terms of executive time, or finance), to act at Levels 1 or III.

Having set out the background of elite interest representation in the European Union, Chapter 2, *The Changes to Treaty Provisions and the Development of DGIV* examines developments in the responsibilities of the European Commission. The Chapter initially explores the ramifications of the Treaty of Amsterdam in air transport policy, in particular its probable effect on the slot-allocation Regulation, the delay to which is a fundamental feature of this research. The effect of the implementation of the Co-decision procedure on air transport policy as a result of Amsterdam (now ratified), is an area that even the most recent literature, in addition to those practising air transport interest representation, have tended to overlook.

The bulk of Chapter 2 is, however, a highly detailed examination of the Competition Directorate of the European Commission, DGIV. This Directorate, together with DGVII - the Transport Directorate - played a crucial role in the examined case studies. Chapter 2 explores the nature of DGIV as a unique, *quasi- judicial* institution of the European Commission. The Chapter further examines DGIV's role in the prevention of cartels, the regulation of state aids to industry, the prevention of abuse of a dominant market position, and its ability to vet and block mergers and industrial take-overs. Chapter 2 also examines the legal basis for the operations of DGIV and the methodology by which it can initiate proceedings. Such an examination is crucial if one to comprehend the unique *quasi-judicial* powers of DGIV. It is necessary to comprehend how these powers may affect the Proposal-making dynamics of the Commission, and how they may affect interest group representation and ultimately the outcome of the two case studies. The unique competencies of DGIV may well act as a buffer to the notion of elite interest representation examined in Chapter 1. If this is the case then it may be that the principle of liberal intergovernmentalism, or even its modified form, neo-liberal intergovernmentalism, may be diluted as a result of *quasi-judicial* factors.

Having examined the theoretical background to elite interest representation, and the background to the *quasi-judicial* institution which may dilute such representation, Chapter 3, *The Legislative Background to EU Air Transport Policy* will examine the legislative background to air transport liberalisation in the EU. This is necessary in order to grasp an empirical understanding of the case studies of ground-handling and slot-allocation, as well as the issues surrounding them. The Chapter summarises the legal challenges to the historical non-compliance of air transport policy in the European Community to the competition provisions of the Treaty of Rome, introduces the primary Court case which determined that air transport was indeed covered by these competition provisions and introduces the subsequent 'three packages' of air transport liberalisation. The Chapter finally introduces the existing legislation concerning ground-handling and slot-allocation that followed the three packages.

Chapter 4, *EU Air Transport Interest Groups*, is a detailed analysis of the groups which are to be examined on their relative successes in attempting to achieve particular policy goals in the two case studies. The primary interest groups examined are the Association of European Airlines (AEA), the European Regional Airlines Association (ERA), and the airports group, Airports Council International (ACI).

The AEA consists almost entirely of what can be classified as elite airline interests, i.e. the airlines which are, or once were in state ownership, and have retained their 'national champion' eliteness. ERA consists of airlines that are, in the main, fully privately owned. The financial turnover of these airlines and the number of personnel they employ are a small fraction of that of the AEA airlines, and subsequently members of ERA cannot, in the main, be considered to be elite interests. Having established this, the Chapter details the way in which elite interests from the AEA have begun to control a relatively small number of the airlines within ERA (for example, the Dutch flag carrier airline KLM recently acquired a 100% stake in the regional airline Air UK, now KLM UK). Although these airlines comprise a relatively small percentage of its members, the evidence suggests that ERA has recently begun to give greater emphasis to the views of this minority of elite owned members within, what is still primarily, a non-elite group.

The third group to be examined is the airports group ACI. Unlike the groups representing airlines, ACI is a 'non-exclusionary' group that represents both elite and non-elite interests. However, although it is the case that ACI consists of over 180 members, the evidence strongly suggests that there is a small group of elites within ACI which have the greatest amount of influence on its position formulation.

This Chapter also looks at two interests that do not fit into the criteria of these groups in order that an even-handed comparison can be made. One is a UK airline (Virgin Atlantic), which, because of its nature does not fit into the AEA or ERA, the second, an 'independent' ground-handling firm (Servisair).

Chapter 5, *The Positions of Interest Groups on Slot-allocation and Ground-Handling*, examines in detail the positions of the AEA, ERA, ACI as well as the views of Virgin Atlantic and Servisair. Furthermore, Chapter 5 examines the way in which the groups arrived at their various positions, the differences of the positions of the elite and non-elite interests, and the adherence of the particular groups to the positions of larger firms within their own membership.

Chapter 6, *Commission Response to the Interest Representation of the Air Transport Industry*, is an examination of the outcome of the two case studies, focusing primarily upon the responses by DGVII (transport), and DGIV (competition) of the Commission to the various interests (examined in Chapter 4), and their demands (examined in Chapter 5). The Chapter therefore ascertains whether a neo-liberal intergovernmentalism can be applied in the case studies (hypothesis number 1). It subsequently examines the way in which DGIV may be able to utilise its wide ranging set of powers (examined in Chapter 2), to circumnavigate the demands of elite interests in order to achieve its own particular goal - that of furthering competition in the EU air transport market. Furthermore, the examination of the case studies ascertains whether it is the situation that the necessity of domination by elites, in addition to effective position making within interest groups, is a necessary prerequisite for a group's effective representation to policy makers (hypothesis number 2). The key test is whether this second hypothesis can be equally applied to the *quasi-judicial* element of the European Commission, or whether this practice affects only the position of DGVII, a legislative element. Chapter 6 will be followed by the Conclusion, which will draw together the theoretical, and empirical information contained within this book. It will also attempt to draw lessons for the air transport industry, in particular, the UK air transport industry, for its future interest representation in Brussels.

Notes

1 The text of the Chicago Convention can be found in the United Kingdom Command Papers, No.6614, 1945. A verbatim text can also be found in: Appendix A (Part 2), in Peter Martin and Elizabeth Martin (Shawcross and Beaumont Solicitors) (1998) *Air Law*, Fourth Edition, Vol.2 (London: Butterworths, London, 1998), pp.A91-A118. For further information on IATA, see: (Web site) http://www.iata.org.

2 See: Hussein Kassim, 'Air Transport Champions: Still Carrying the Flag', in Jack Hayward (eds) *Industrial Enterprise and European Integration - From National to International Champions in Western Europe* (Oxford: Oxford University Press, 1995), pp.189-214.

3 The exception was the US domestic air transport market which is widely considered to be the first air transport market to have been 'de-regulated'. This did not happen however until 1986. A discussion of the US market and its applicability to the EU market can be found in, 'US Airline Deregulation: The Lessons for the European Community' (UK Dept. of Transport, Working Paper No.109, Sept, 1989). See also: Greg Williams, *The*

Airline Industry and the Impact of Deregulation (Aldershot: Avebury, 1994); Alan P. Dobson, *Flying in the Face of Competition* (Aldershot: Avebury, 1995).

4 The notable exception to this rule was the creation of the airline SAS (Scandinavian Airlines System), in 1946. 50% of the shares of SAS have been owned, since its establishment, by the governments of Denmark, Norway and Sweden; 50% is privately owned (source: *Flight International,* Vol.149, No.4516 (27 Mar-2 Apr, 1996), p.92). The UK flag carrier, British Airways was not formed until 1972 upon the amalgamation of two state owned airlines, British European Airways (BEA), and British Overseas Airways Corporation (BOAC). For further information, see: Alison Corke, *British Airways: The Path to Profitability* (London: Pinter, 1986).

5 See: Articles 85 and 86 of the Treaty; see Chapter 2 for further details.

6 Having said this, in the 1970s the UK government were determined to protect BA's position in the air transport market, even from domestic competition. This led to the infamous 1977 High Court case where the UK's government's attempt to prevent Freddie Laker's Skytrain service collapsed after a ruling by Lord Denning, at which he stated:

> Mr Laker is a man of enterprise. He has an exciting project for travel by air. He wishes to start a service across the North Atlantic from England to the United States. It is quite unlike conventional air services. It is to be more like a railway service. Passengers are not to reserve their seats in advance. They are to go to the airport, buy their tickets and board the aircraft – all in one sequence – just as passengers go to a railway station and buy their tickets and join the train. So Mr Laker calls his project Skytrain. He hopes to attract a hitherto untapped source of traffic. He is going to cut out all the travel agents and their expensive commissions. He is going to charge fares much lower than other carriers. He is not going to fly from Heathrow, but from Stansted; so there are no in-transit facilities. By these measures he suggests that he will not take away traffic from the established airlines, but will create new custom. He will carry passengers who would not otherwise have travelled at all...

and so Lord Denning continued (quoted from: Howard Banks, *The Rise and Fall of Freddie Laker* (London: Faber & Faber, 1982), p.31.

7 Joined Cases 209-213/84, Ministère *Public v. Lucas Asjes:* [1986] E.C.R. 1425, [1986] 3 *Common Market Law Review* 173. See Chapter 3, for further details.

8 Case 66/86, *Ahmed Saeed Flugreisen and Silver Line Reiseburo GmbH* v. *Zentrale zur Bekämpfung unlauteren Wettbewerbs* E.V.: [1989] E.C.R. 803, [1990] 4 *Common Market Law Review* 102.

9 This process is described as 'negative integration'. For further information, see: J. Pinder, 'Positive Integration and Negative Integration: Some Problems of Economic Union in the EEC', in *The World Today* No. 4 (1968), pp.88-110.

10 See Chapter 3.

11 See Chapter 6.

12 See Chapter 2.

13 See: David A. Baldwin (eds), *Neo-realism and Neo-liberalism - The Contemporary Debate* (New York: Columbia University Press, 1993).

14 See also: Carole Webb, 'Introduction: Variations on a Theoretical Theme', in Helen Wallace, William Wallace, Carole Webb (eds), *Policy Making in the European Communities* (Chichester: John Wiley and Sons, 1977), p.9.

15 See, in particular, Leon N. Lindberg and S.A. Scheingold (eds), *Regional Integration: Theory and Research* (Cambridge, Massachusetts: Harvard University Press, 1971).

16 Examined in Chapters 2 and 3.

17 Interview: Brian Simpson MEP (Member of the Transport and Tourism Committee), Brussels, 20.1.97.

18 For furthermore information on the integration of EU air transport regulatory authorities, see, paper presented by Yves Morier (Regulations Director), Joint Aviation Authorities, 'The Joint Aviation Authorities (JAA) Safety Strategy Initiative', paper presented at IBC (London), 'Aviation Safety Management' conference, London, 15.5.98.

19 Tony Edward (Chief Executive), Messers Dowty Aerospace, 'Commercial Aspects of Aerospace', paper presented at the Royal Aeronautical Society, 'A Rough Guide to Aerospace', conference, London, 18.3.98.

20 The UK Office of Fair Trading / Monopolies and Mergers Commission (now the UK Competition Commission).

21 'Commission Proposal for a Council Directive on access to the ground-handling market at Community airports', COM (94) 590 final, 13.12.94.

22 'Council Regulation 95/93 on common rules for the allocation of slots at Community airports', O.J. L 14/1, 12.1.93.

23 Ibid, p.6.

1 The Theoretical and Historical Considerations of Interest Representation

This Chapter is divided into three Sections.

Section 1 will review the theory of liberal intergovernmentalism. This is in order that a modified version of the theory, *neo-liberal intergovernmentalism,* can be applied to the workings of the European Commission in the case studies of the formulation of a Commission Proposal for a Council Directive on airport ground-handling, and the attempted formulation of a Commission Proposal on slot-allocation. These case studies are, essentially, examinations of elite and non-elite air transport interest representation. Section 1 also examines 'two level game theory', 'elite accommodation theory', and 'reverse state strength' theory, all of which are applied to the two case studies.

Section 2 will establish working definitions of 'elite' and 'non-elite' interests. Such definitions are necessary for accurate examinations of the relative successes of elite and non-elite air transport interests. Subsequently, Sections 1 and 2 set out the background for the application of Hypothesis Number 1.

Section 3 on the other hand examines theories of interest group decision making to provide the background to Hypothesis Number 2. Primarily, this section examines the theory that elites may not only dominate policy formulation within the state, but may also dominate the behaviour and actions of interest groups. Section 3 also examines how elite interests may manipulate interest group representation for their own advantage. Such 'interest group manipulation' features prominently in the examined case studies.

1. Liberal intergovernmentalism - extending the boundaries of intergovernmentalism

As discussed in the Introduction, liberal intergovernmentalism incorporates aspects of *neo-functionalism.* Neo-functionalists maintain that states can no longer be considered to be the central actors in the international system. Furthermore, trans-national policy making bodies, such as the European Commission and European Parliament, have shifted the focus of decision making away from individual states. Subsequently, states are no longer seen to be some form of ultimate policy making unit (Haas, 1961, pp.366-392; 1971, p.6; Nye, 1968; Puchala, 1972, pp.267-285; Schmitter and Streeck, 1994, pp.169-187). Qualified Majority Voting (QMV), at the Council of Ministers is an example of the diminishing power of the state. As Sbragia argues:

> With QMV each state needs to search for allies, whether or not it likes the Commission's Proposals....The Member-States, as a collectivity, are still central to the decision-making process, but not all the states will necessarily be central all the time. The veto made each one central, while the use of majorities changes the focus to coalitions and away from states acting individually.[1]

On the other hand, liberal intergovernmentalism ascertains that it is the 'elite actors' within a state which substantially influence a state's position on a particular issue. The state subsequently represents the elite actors at the inter-state, or (as in the case of the EU), the inter-Member State level.

Moravcsik argues that whilst the state is of prime importance in international negotiations, it is the domestic interests within the state that influence its negotiating position. Liberal intergovernmentalism encapsulates both 'national preference' theories (which originate in the intergovernmentalist school), and interdependence theories (which originate in the functionalist / neo-functionalist school).[2]

Interstate policy goals are defined by domestic criteria and are a response to shifting pressures from domestic societal groups whose preferences are channelled through political institutions. The key determinants of the goals of the state are the identity of its important societal groups (or 'elite' interests), the nature of their interests and their relative influence on domestic policy, argues Moravcsik.[3] For example, in the course of the bargaining that led to the Single European Act, Member Sates formulated their approach according to their own domestic policy concerns. The net result was a 'trade off' of preferences between the Member States, as they attempted to satisfy their own domestic interests.[4]

'Two level game' theory

Like Moravcsik, Putnam claims that it is the domestic interests that influence a state's position on a particular issue, as domestic interests must be appeased for a government to retain political power.

> At the international level, governments seek to maximise their own ability to satisfy domestic pressures, while minimising the adverse consequences of foreign policy developments.[5]

However, Putnam also promotes the theory that inter-state negotiation is a 'two level game'. Level 1 is the negotiation between states on a particular issue, for example, the negotiation between the Member States of the European Union on air transport. Level 2 is the contact between the state and its domestic interests concerning the particular issue, for example between the French government and Air France.

Therefore, the domestic game (Level 2), may determine the range of possible agreements between states that can be ratified; this is described by Putnam as a 'win set'. Level 2 can also be utilised by the state in order to achieve the desired outcome at Level 1, often by the 'controlled exchange of partial information'. Subsequently, domestic pressures (Level 2 pressures), may well be 'exaggerated' by the state (acting at Level 1), in order to accomplish this.[6]

'Elite accommodation' theory

Hix argues that a 'cartel of elites' exists in Member States, where certain actors maintain an increased importance in relation to other actors. A government must satisfy this cartel if it is to maintain its own political survival. Such a situation is described as an 'elite accommodation'; the views of the elites carry a far higher proportion of influence than those outside this cartel.[7] Hix also argues that comparative politics present the most relevant framework for the study of EU politics and bargain making, as opposed to 'traditional international relations approaches'.[8] Hix criticises integration theories for their concentration on the 'formality and objectivity of integration' (such as EC law), and states that comparisons should be made with models of domestic government. Hix argues that the political relevance of an issue (i.e. the possible effect a particular policy may have on the popularity of the Member States' governments), presents a more accurate indicator of EC policy making, than what may be generally considered to be in the 'national interests' of Member States.[9]

Reverse, 'state strength', negotiation theory

Huelshoff argues that interests can be 'Co-operative', or 'Competitive' in a policy area.[10] The former category, as the terminology indicates, are interests that are co-operative in that their views do not oppose one another. However, the interests may not necessarily be 'working together'. For example, the views of British Airways and the major UK airport operator, BAA, on 'slot-allocation' are very similar. Therefore the two firms are Co-operative despite the fact they may not be jointly presenting their views to policy formulators. However, the views of London City Airport and British Airways on slot-allocation are diametrically opposite; therefore, their interests are Competitive.[11] In Chapter 5 we find that the German airport interests in regard to the issue of ground-handling were highly Co-operative, giving them a particular advantage in negotiations.

Whether interests are Co-operative or Competitive plays a deterministic role on the amount of effective pressure that they can apply to their government. Co-operative interest representation results in the state being in an increased position of vulnerability at the domestic level. This domestic vulnerability is reversed into strength at the inter-state level of negotiation, as the state has less 'room for manoeuvre'. The weakness of the state domestically (because of the strong position of Co-operative interests), results in a reduced amount of flexibility for the state when in inter-state negotiations. Subsequently, the state becomes 'locked' into a position (resulting from the relatively strong domestic interest representation), and cannot easily alter its bargaining position. Conversely, if its domestic interests are Competitive, the state will be *less* vulnerable at the national level. However, at the inter-state level the state will have greater flexibility in its negotiations. This flexibility results in the state being in a *weaker* position in relation to the other states.

There also exist additional factors which may influence a state's weaknesses or strengths in inter-state policy negotiation, argues Huelshoff.[12] States with centralised institutions and majority governments are characterised as strong states at the national level. States that are federations, those with a tradition of an independent judiciary and minority governments are categorised as weak at the national level. Because of their respective institutional compositions, weak states are more vulnerable than strong states to influence from their domestic interests. As a result, weak states are usually in a relatively stronger bargaining position at the Council of Ministers than states that are strong at the domestic level.[13] Utilising Huelshoff's definition, the UK is classified as a strong state (in relation to interest groups), whilst Germany (a federation), is weaker in its relationship with interest groups. If correct, it would be the case that at the Council of Ministers the UK is in a relatively weak position to make its wishes prevail than Germany (as the bargaining strength is reversed at the inter-Member State level).[14]

Table 1.1 Summary of Huelshoff's theory

Domestic Interests		**Domestic Government**		**Negotiating Position of State**
Co-operative	⇒	Weak	⇒	Strong
Co-operative	⇒	Strong	⇒	Moderate
Competitive	⇒	Weak	⇒	Moderate
Competitive	⇒	Strong	⇒	Weak

Summary of liberal intergovernmentalist arguments

The liberal intergovernmentalist argues that an EU Member State's, bargaining behaviour changes according to the perceived result of accommodating various domestic interests. This suggests a more complex and dynamic set of political benefit / cost calculations than has traditionally been considered. Lindberg's assertion, 'getting the politics right may take precedence over getting the economics right',[15] may result in a situation where state policy formulation is largely a political calculation. The above theories are based upon the assumption that domestic policy is of prime importance in the formulation of EC policy. Subsequently, the theories are a challenge to the neo-functionalist tradition. However, the examinations in which Moravcsik defined *liberal intergovernmentalism* were examinations of inter-Member State negotiations leading to the formulation of the Single European Act and the Treaty on European Union.[16] Both pieces of policy formulation required unanimity by the Member States of the EU. Moravcsik has not necessarily shown whether his theory would work in a Qualified Majority Vote situation.

The case studies contained within this book attempt to ascertain if the liberal intergovernmentalism described by Moravcsik is applicable to policy Proposals that were, and are, ultimately to be decided by Qualified Majority Vote. The case studies also reflect upon elite accommodation and differing effects of Co-operative and Competitive interests. In sum, the key test is whether the properties that comprise liberal intergovernmentalism extend to the actions of the European Commission, in a *neo-liberal intergovernmentalism* form.

2. Elite and non-elite interests

Domestically, firms strive to achieve a situation where business risk and uncertainty are reduced by the establishment of a relationship with the policy formulator.[17] This 'traditional national approach' is thought to have provided effective access, via the host state, to the Council of Ministers for the formulation of EC legislation (Greenwood, Grote, Ronit, 1992, p.8).[18] Certain firms have been relatively successful in establishing such relationships with policy formulators in order to reduce business uncertainty. Such firms are described by Moravcsik *et al* as 'elites'. However, although hypothesising on the role of elites, Moravcsik *et al* do not adequately provide a comprehensive definition of an elite. Furthermore, the case studies within this book examine 'interest groups' which represent both elite and non-elite firms. It is therefore essential, prior to such an examination, to define the 'properties of eliteness'.

For the purposes of the case studies contained within this book, elites are classified as those firms to which the greatest political importance is attached by their host governments. Ordinarily, these firms are the 'largest' (see below), airlines or airports within a particular Member State of the EU.[19] These firms are also classified as 'national champions'.

Elites and 'National Champions' – The background

Elite firms, or national champions, are the firms which have been the most successful in establishing relationships between themselves and policy formulators (Sandholtz 1992, pp.113-142).

The UK government, in the aftermath of World War 1, formed many firms in order to strengthen its hold on the British Empire. Imperial Airways (later to become the British Overseas Airways Corporation, and then British Airways), was formed in 1924[20] on the basis that it was to be a state subsidised carrier on routes to the British Empire (Turner, 1984, pp.12-13). Its creation was made possible by a network of business people and politicians who co-operated to control the state intervention in this, and other sectors of the economy. Resulting from successive governments' lack of knowledge of air transport, the management of Imperial Airways was literally able to dictate to the government its policy on air transport (Reader, 1977). This situation was also reflected in the UK oil industry, post WW1. Government policy was frequently based on myths and incomplete information created by what had become the 'national champion' firm. For example, the Anglo Persian Oil Company (APOC) is alleged to have aimed anti-Semitic propaganda at Royal Dutch Shell.[21] Because of the government's absolute reliance on APOC it acted upon the information it received and felt there was no need to question it. The examples of Imperial Airways and APOC were reflected across Europe in an array of industries. Subsequently there was a propensity by European states to discriminate in favour of their particular national champions; this was also the case in Japan, and to a lesser extent, the United States (Stoppford and Strange, 1991, p.57).[22] Furthermore, the co-operation of the national champion firm has often been considered essential for successful policy implementation (Grant, 1993, pp.46-47).[23] As a result, policy formulation and implementation have become inseparable.[24]

Therefore, a national champion has been, or is, a state owned firm. The firm has traditionally had a monopoly in that market which was often kept artificially in place by its host government.

From the mid-1980s many national champion firms in Europe began to move from public to private ownership. This was partly as a result of the fact that the market sectors in which the national champions operated (for example telecommunications or air transport), were being liberalised in order that competition between firms could take place.[25] However, it may be argued that privatisation, far from distancing national champions from the policy formulators of their states, has allowed a situation to continue where the national champions retain their influence over their particular governments. Although theoretically operating in a free market, their status of national champions, and hence the propensity by states to discriminate in favour of their particular national champions (hence the 'eliteness' of the firm), has largely remained.[26] An example is the car manufacturer Renault, which enjoys a status in France which *de facto* guarantees it its financial solvency (McLaughlin, Jordan, Maloney, 1993, p.197). This link between the national champion (or elite firm), has remained because of the political importance that the firm continues to carry. Because of the protection from competition provided by states for national champions, and because of the large amount of capital injected into these firms (usually over a period of several decades), national champions (whether in state ownership, or more recently private ownership), are normally the largest nationally based firm in a particular industry, with the greatest financial turnover, and employing the greatest number of workers. Such firms will subsequently possess a high degree of *political* as well as economic importance in their particular state. Issues affecting employment within these industries carry a particularly high degree of political significance. Elite firms, because of their protected background, often employ substantially more people in relation to the size of the firm in comparison to non-elite firms.[27] Therefore, national champions remain elites as long as they carry the political importance that is attached to them. Conversely, non-elites (or non-national champions), are normally firms which are fully privately owned and that are also in competition with elite, national champion, firms. Often the existence of such firms has only come about as a result of liberalisation in that particular sector. Such firms have not enjoyed the protection and subsidies that have been received by their elite (national champion), competitors. As a result, the financial turnovers of such firms tends to be significantly less than elite firms. However, it is often the case that the profitability of non-elite firms is greater than elites, resulting from more efficient organisation.

National champions and EC legislation

Sandholtz points to the deteriorating importance of the 'national champion strategy' as a result of the competition arising from 'greater economies of scale'. Sandholtz's case study ascertains that the national champion strategy in Europe, in the field of the electronics industry, was insufficient to compete with the competition emanating from America and Japan. To face such competition, an integration of sectors of industry took place. Sandholtz concludes that interest representation at the national level became less effective as the companies concerned had interests that transcended the boundaries of the Member States of the EU; it therefore became impractical for one government, acting as a unitary actor, to represent its national champions, and a 'pan-EU' approach needed to be implemented. Similarly, a study of the firm Airbus Industrie ascertained that the cost of manufacturing a new airliner was increasingly beyond the capacity of a single EU Member State (Sandholtz, 1992, pp.113-142, esp. pp.118-119).[28] The aerospace industry is one where cross-ownership and collaboration increasingly exist across the EU and therefore national representation is reduced in its relevance and in its effectiveness. The chemical industry is a further example of a national champion industry that has traditionally been represented at the highest levels by its host government. However, similar problems in representation also faced this industry. In the 1980s the UK industrial chemical industry pressed for national regulation, whilst the German chemical industry pressed for Community regulation. Therefore, the same industry produced two different nationally based views on the issue.[29] In such a situation where the national champions of two Member States disagree with one another over an aspect of potential Community policy, the national champions can no longer rely solely on their Member State, and Community coalitions need to be formulated in order to be effectively influence EU policy making.

The examples of electronics, aerospace and chemicals may suggest that the importance of the individual Member State as a target of representation is reducing. However, the elite firms which have traditionally dominated policy formulation at the national level may well retain their importance in determining EC policy formulation. If this is indeed the situation then an application of *neo-liberal intergovernmentalism* may have occurred; i.e. the focus has shifted from a national champion - state relationship to national champions (plural) - EC policy formulators relationship.

Stephen George claims that interests, such as individuals, firms or interest groups, bring certain resources into policy formulation. These resources include informational, organisational, authoritative and political resources.[30] Informational resources allow the Commission to avoid formulating a policy that has no relevance or which may be impossible to administer. The resources of authority and organisation allow the interest the opportunity to influence policy formulation, because the interest will have demonstrated that it has the 'know how' to help draft policy and that its co-operation is necessary for the policy to be implemented successfully. The political resource is the fact that the interest, in the case of elite firms, will usually carry the support of the Member States - essential if the Proposal is to achieve a Qualified Majority or unanimity (if needed) in the Council. It is, primarily, the national champions within each Member State that possess the ability to engage all four resources in an effective relationship with the Commission. It is, ordinarily, the national champions, or the elites, which can provide the authority, organisation, and the political viability to allow a policy to be successfully formulated. Whilst many firms and interest groups may be granted *access* to policy formulators, relatively few have any significant *influence* over the policy outcomes.[31]

3. The relevance of groups - A reflection of liberal intergovernmental theory

Having discussed the properties of liberal intergovernmentalism and the properties of eliteness, let us now discuss the proposition that an 'elite accommodation' may exist, not only within the state, but also within interest groups.

A useful place to start in the discussion of such a proposition is Averyt's 'orthodox strategy', first mooted in 1977. The orthodox strategy ascertains that firms may follow a 'dual track' in their representation in the European Union.

Track 1 The firm may apply representation to the relevant government department at the national level, either as a unitary actor or via a national interest group. The relevant government department then represents the firm at the Council of Ministers.

Track 2 The firm may be represented by a national interest group, which is subsequently represented by a 'Euro Group', or 'Peak Association' (an association of nationally based interest groups),[32] within the EC policy making framework (Averyt, 1977, p.102).[33] However, the 'orthodox strategy' presents two interrelated dilemmas. Track 1 potentially presents the problem of Member State representation put forward by the neo-functionalists, namely that many issues transcend national boundaries, and cannot be effectively dealt with by the representation of a single Member State (particularly if the issue is one dealt with by QMV). Track 2 potentially presents the problem of interest group division; peak associations may have to contend with up to fifteen different national perspectives on a particular issue.[34]

Divisions between firms may not only arise within peak associations but within all interest groups which have a number of contending points of view to represent. Arguably, such divisions reduce the effectiveness of the interest group. As a result, interest groups may be viewed in some quarters as weak, fragmented and unable to agree on common positions. Because of this, interest groups may also be accused of being slow to react to policy Proposals.[35] Firms' conflicting concerns within an interest group may therefore result in a lack of a *proactive common denominator* of interest representation. The development of stable agendas and processes in relation to a policy area is subsequently undermined. The net result is that the Commission must 'choose' from many firms across the EU, as a result of such division.[36]

Olson puts forward the view that interest group membership is largely an unnecessary expense. Olson's hypothesis is that firms gain greater relative advantage by 'free-riding' the activities of interest groups which have similar aims (Olson, 1965). This hypothesis rests upon the basis that the costs of pursuing policy as a group are substantial, and thus free riding (where an interest benefits from the representation of a group of similar firms, without being a member *de jure*), is considered the most beneficial option.[37] A case study by Jordan / McLaughlin of the representation of the car industry deduced that interest groups were indeed fragmented in their representation and inefficiently organised, often resulting from 'intra-organisational political difficulties.' The study ascertained that, 'Groups often become paralysed because they cannot maintain consensus amongst conflicting interests'.[38] However, the study also ascertained that individual firms did not benefit from Olson's 'free-riding' as the interest group could not be relied upon to seek acceptable benefits without 'inside steering'. Therefore, interest groups, although far from ideal, were of proportionately greater benefit than a solo pursuit by an individual firm. Free riding is a concept that is of a 'pure' rationality. In practise a 'bounded' rationality exists where companies seek to minimise business uncertainty in 'negotiated environments'.[39]

An example of an interest group that has, to a great degree, effectively dealt with the problem of such fragmentation is the 'Association des Constructeurs Européens d'Automobiles' (AEAC).[40] A study examined AEAC shortly after it had been amalgamated from two peak associations.[41] AEAC significantly altered the decision making structure that it inherited, and its decisions were now to be formulated with a 'qualified majority vote' (requiring a 75% threshold), as opposed to the traditional unanimity needed for its decision making. Additionally, national interest groups were to play a relatively minor role in the decision making of AEAC, thus helping to avoid a situation where each national interest group focused upon its own specific national demands, thereby producing roughly a dozen alternative, nationally based opinions. After this reform, the European Commission noted that the European car industry had begun to, 'speak with one voice'. A particular example of this previous inability to speak with one voice was the European car representation regarding the issue of Japanese car imports. The issue was one where there was a broad diversification of views based upon separate national opinions. This resulted in weak and fragmented representation to the Commission, but also by the Commission in its negotiations with Japan, thus weakening the European car industry *en masse*.[42] Partly as a result, AEAC was formed, and its interest representation began to no-longer resemble a *lowest common denominator* of interest representation, but a *proactive common denominator*. Such division, which was evident prior to the establishment of AEAC is noticeable within the interest group the Association of European Airlines, examined in later Chapters. Such interest groups may do well to look closely at the structure of AEAC and the lessons in interest representation that the car industry has now learnt.

Multiple strategies of interest representation

Some firms are able to adopt *multiple strategies* of interest representation, which consist of utilising, 'multiple points of access' (Streeck, Schmitter, 1991, pp.133-164; McLaughlin, Jordan, Maloney, 1993, p.195). Firms can apply representation directly as a unilateral actor. Additionally, interest group membership grants those firms 'legitimisation'. The view of an interest group promotes the impression that it is the view of a particular industry that is being aired, rather than the voice of an individual concern. However, unilateral representation allows 'one to one' and personal contacts to be established between the firm and the policy formulator. When the two approaches of unilateral and group representation are applied, the firm can be described as utilising 'multiple channels' of interest representation.[43] For example, car manufacturers attach major significance to their 'in house' policy operations in Brussels. They see their activities within the *collective* policy participants (interest groups), as acting in parallel with the *non-collective* policy participation (unilateral action).[44]

However, unilateral action may also have negative side-effects. Despite the overhaul of the representation of the European car industry, and the undoubted lessons that can be learnt by other sectors of industry in this regard, AEAC maintained some of the negative attributes attributed to its predecessors. This weakness resulted from the, 'diverse political tactics of large companies that undermine Euro-groups...'[45] This weakness can be analysed in a number of ways.

1. Member firms may have been willing to reform the structure of the interest group to make the group more effective, but they may be unwilling to sacrifice or limit their independent activity. Therefore, firms will

increase their unilateral interest representation activities when the group fails to represent their particular interests.

2. The most effective interest group representation is when interests resist 'breaking from the pack' (either overtly or covertly), for the benefit of the group as a whole.

In order to achieve these two objectives it is seen in some quarters as necessary to give an increased importance to the membership of large, elite firms within interest groups. A study by Greenwood of the pharmaceutical industry seems to confirm this.[46] The study deduced that interest groups, rather than a collection of national concerns, are of prime importance in the formulation of EC policy.[47] Thus, the groups, acting as pan-EU actors (as opposed to a peak associations acting on behalf of nationally based interest groups), are highly effective in promoting the interests of their members. The study also dismissed Averyt's orthodox strategy because of its concentration on the national approach (Averyt, 1977),[48] and ascertained that the examined interest groups were coherent and effective groups,

> ...which have demonstrated their ability to organise their membership to mutual benefit, and to develop 'clientilistic' relationships with relevant parts of the Commission.[49]

The study deduced that the shifting of the representation of interests from the state has taken place, primarily, for two reasons;

1. The enhanced competencies of Community institutions in the policy area.

2. Because the EU interest groups have developed a 'track record' with the EC policy formulators. Importantly, Greenwood states that the success of this representation has primarily come about because:

> ...the industry has learnt to weigh the interests of the large players against the small. Without the scales being tipped in favour of the large players, a federation is doomed.[50]

Thus, the key element of the success of the interest group is the dominance and subsequent representation of the elite interests.[51] The elite interests are able to decide the direction and nature of the group's approach to Community policy formulators. The views of relatively small members are outweighed by these large players, yet the small players provide legitimacy for the large in that the large players may argue that the view of the group represents a pan-EU industry consensus (see below).

In a study of the European Association of Consumer Electronics Manufacturers (EACEM),[52] Cawson noted that EACEM, in terms of its *actions* was relatively insignificant, yet its role as an *agent of legitimisation* for the views of member firms was invaluable.[53] A study by Schneider of the telecommunications industry also revealed that access to the influential channels of the Commission is rarely equal. Schneider found that for an interest to most effectively wield influence it must be 'recognised' by the Commission in order to be invited onto an advisory committee or if it is to be regularly consulted by the appropriate divisions or *services* within the Commission (Schneider, 1992, pp.42-68).[54] Schneider conducted a study of the formulation of a 1987 Commission Green Paper, which was aimed at linking the telecommunications market with the 1992 internal market goals.[55] Individual corporations (elite, national champions) were the only actors that played an important influence in the formulation of the Green Paper. Schneider therefore argued that the Commission possessed a 'hierarchy of actors'. He deduced that the key feature of corporatism[56] - the ability of interest groups to enter into bargaining commitments - is still lacking. The reason, Schneider believes, for this lack of a 'Eurocorporatism' can be found in a recent information bulletin from the Union of Industrial and Employers Confederations of Europe (UNICE).

> European legislators are likely to receive so many diverging messages from so many different interest groups that they could end up either very confused or able to divide and rule...EC-representations of business and industry must be more aware of each other...Their messages need not be identical, but should at least be on parallel tracks. If and when they feel it necessary to contradict the views expressed by other bodies, they should do this only after fullest consideration and genuine attempts to reconcile the differences prior to going public.[57]

As mentioned above, the 'legitimacy' of a group consisting primarily of elite firms may be increased by the membership of relatively small firms in the interest group. (As Chapter 4 details, the Association of European Airlines has allowed the airline British Midland to join its ranks, despite the fact that its membership consists overwhelmingly of flag carrier, elite interests.) Elite firms may sometimes subsidise the membership of non-elite firms within 'their' group.[58] Such a subsidy can be considered an investment in the effectiveness of the interest representation of the interest group. However, it is often the firms or interest groups that establish themselves as the most effective providers of information that are most valued by the Commission and other EC institutions. There subsequently exists a process of *interest representation establishment,* where the Commission will become 'familiar' with various interests as reliable and dependable sources of information.[59] The Commission seeks accurate information for policy formulation; elite firms possess greater resources and greater access to this information, and they can subsequently supply this information to the Commission. Chapter 5 details the way in which the interest group Airports Council International made its self the primary and most effective source of information to the Commission when the Commission was drafting the Proposal for the ground-handling Directive.

Schneider and Cawson identified the *inability* of groups to present a single coherent message as the reason for their failure in influencing the EC policy formulation process. In Greenwood's study the interest group presented a single coherent message, and was therefore able to influence the EC institutions. Crucially important is that in Greenwood's case-study, the views of the elite firms were dominant *within* the group, allowing the interest group to be relatively efficient. In the case-studies of Schneider and Cawson the interest groups did not take appropriate account of the views of its elite members; thus the elite members 'broke from the pack', when such firms felt that their views were not adequately represented. The difference between the studies is that in Greenwood's case the interest group was effective because the influence of elites within the group was proportional to those firms' size and political importance.

A reflection of the theory with reference to airlines

There have, until comparatively recently, been few examinations of the role of air transport interests and representation regarding EC policy formulation. This is primarily because the issue has traditionally been one of a 'state centric' nature. However, a 1992 study examined the role of the European airline flag carriers in their lobbying to amend a Commission Proposal for a Council Directive (Van den Polder, 1994, pp.103-119). The Proposal concerned Computer Reservation Systems (CRS's), and was initially opposed only by the Dutch airline KLM. The Proposal stipulated that where 'code sharing agreements'[60] existed, the two airlines involved in the code share must not be shown together on the same CRS screen.[61] (KLM had recently established a code sharing agreement with the American airline Northwest). The Dutch government supported KLM's opposition to the Proposal, and the Dutch government subsequently presented its views at a meeting of the Council of Ministers in February 1993. However, the Netherlands had little support at that meeting, and no support from the other flag carriers of the European Union. KLM and the Dutch government were unable to rectify the Proposal at this stage. However, once KLM had won the support of the other flag carrier airlines (except SAS of Denmark and British Airways), the Member States followed the line taken by their own flag carrier airlines (all members of the group the Association of European Airlines),[62] and agreed the KLM amendment to the Proposal (other than the Danish and British delegations, who likewise, followed the interests of their flag carriers). Eventually, Britain and Denmark did in fact drop their objections and in June 1993 the Council of Ministers unanimously adopted the 'KLM amendment.'

Van den Polder puts forward two main observations resulting from this case study.

1. It was the dominant (or elite) interests, within the Member States, that were instrumental in altering the positions of their member governments. Van den Polder points to the fact that most of the European 'flag carriers' possessed close links with their governments and this close relationship was highly beneficial in the pursuit of their cause. He contrasts this with the 'looser' relationship that existed between British Airways and its government, as a result of British Airways being in fully private ownership;

this aspect differentiated British Airways substantially from the majority of other European 'flag carriers' at the time.[63] Furthermore, this case study appears to show that the views of Member States can change substantially if they are presented with a firm, near single view (co-operative interest representation), from a broad coalition of relatively large European interests. It is clear from this study that relatively large airlines formed the major interest representation in this matter. Relatively small airlines played, according to the study, no role.

2. Van den Polder recognises that the lobbying by KLM was at a very late stage in the process of policy formulation. He recognises that in order to have the maximum effect on EC policy formulation air transport interests needs to become involved in the process at the earliest possible opportunity.
Van den Polder uses the following analogy:

> If you want to direct the ship into the port of Rotterdam, you have to start applying the rudder, much earlier, somewhere off Ostend. An effective lobby therefore begins at as early a stage as possible, that is to say, within the Commission before the proposal acquires any public status - in other words, the stage at which it is initiated and the proposal is being developed.[64]

Van den Polder therefore supports the 'general consensus' of the interest representation literature of the 'early approach' for successful interest representation, preferably prior to the Commission formulating its Proposal.[65] However, Van den Polder's own case study shows that a Proposal, once formulated, can, in actuality, be altered, although the relevant interests make life far more difficult for themselves. In the CRS case study the views of the Member States reflected precisely the views of their flag carriers. When the views of the flag carriers changed, so too did the views of the Member States. Van den Polder is clear that a goal is best pursued if other interests in that field are also supportive. Indeed, Van den Polder may have under emphasised the importance of this factor in this particular case. The goal of KLM was probably attained because the vast majority of other actors which were elite firms (the flag carrier airlines), eventually supported the stance of KLM and were then able to influence their own particular Member States. KLM's goal could not have been achieved without such support.

As this book utilises case studies of air transport, it is necessary to elaborate on the *raison d'être* of the decision by the majority of European flag carriers to acquiesce to the views of KLM. Importantly, it is highly likely that this situation could only happen in an environment where the decision at the Council of Ministers was taken by a Qualified Majority Vote. In such a situation, it was clearly fruitless for the UK and Denmark to have maintained resistance, as the other Member States could clearly have outvoted them; the UK and Denmark subsequently abstained from the vote. If the Proposal had required unanimity, Britain and Denmark could have extracted a 'price' from the other Member States, or ultimately have vetoed the entire amendment.[66] Although Van den Polder accurately analyses the process of policy formulation, the reasons why the vast majority of the airlines of the AEA changed their positions to favour the KLM amendment must also be mentioned.

At the time, the only European airline in a code sharing alliance was KLM. The Proposal, as it stood, was detrimental to the interests of such airlines. However, since that time, SAS (Scandinavia), Lufthansa (Germany), Air France, Alitalia (Italy), Sabena (Belgium), have all entered alliances.[67] British Airways and Iberia are amongst the latest airlines to attempt to follow this track and will probably be followed by others, dependent upon further EC regulation in this area. Therefore, at the time, KLM would have been put at a disadvantage as a result of the Proposal as it stood. However, KLM was able to convince the other flag carrier airlines that, because of the trends in aviation markets, they would also soon have to enter similar alliances (which indeed happened). KLM was therefore able to persuade the majority of the other European flag carriers that it was in their long term interests to support the amendment. Primarily, the policy formulation process proved to be flexible, and responded to the demands of the *elite interests.*

Other factors affecting elite and non-elite representation

Research indicates that a primary resource lacking in non-elite firms is the 'executive time' needed to monitor policy formulation at both the national and EU levels. Non-elite firms, which usually do not employ the personnel in order to carry out this task (and who may not be able to afford such personnel), are at a substantial disadvantage in comparison to those firms that have the financial resources to employ such personnel. Executives from non-elite firms have suggested that 'lobbying' and 'interest representation' is usually a low priority on their agenda. On the occasions that they do lobby, the process often consists of very brief, sporadic, meetings.[68] Elite firms, however, usually possess a division that deals with governmental or EC matters. They 'monitor' developments in the policy sector, reporting back to the directors of the firm and take action where necessary.[69] All of the elite firms questioned in this research possess a governmental affairs department; this factor is subsequently a key distinguishing characteristic of elites, compared with non-elite firms.[70] Elite firms establish direct relationships and 'every day' contacts at a variety of levels in government. Often members of boards will be engaged in *ad hoc* discussions with ministers, their advisors, and civil servants.

Cawson's analysis of the European electronics' industry found that the examined firms utilised 'multiple points' of influence (Cawson, 1992, pp.99-118). A clear distinction between elites and non-elites in the air transport case studies contained in later chapters is the ability of elites to engage in a 'triple track' of interest representation. Elites apply pressure nationally (Level 1), through their interest group (Level II), and directly at the Commission and European Parliament (Level III). British Airways (listed as an elite interest),[71] displays an example of the importance that elite firms place on interest representation. British Airways employs at least eight executives whose full time task is governmental and legislative related matters; this number includes a full time parliamentary affairs manager, and a representative in Brussels (interestingly only established in 1997). Additionally, four 'lobbying firms' were until recently contracted to represent the opinions of British Airways at Westminster and in Brussels.[72] Similar emphasis in representation is mirrored by elite air transport interests across the EU. Non-elite interests are handicapped at each of the three levels. They do not possess the resources (in terms of finance or personnel), to lobby at the national level (Level 1), or directly in Brussels (Level III). There is therefore a near absolute reliance by non-elite interests on Level II (interest group) representation. This is the case for the majority of the members of the airline interest group the European Regional Airlines Association, examined in Chapter 4. Although interest groups provide a 'legitimacy' to the views of both elite and non-elite interests there are, inevitably, situations where the views of the composite interests differ, resulting in a 'lowest common denominator' of interest representation.

The study of the re-formed AEAC is supported by a US study which ascertained that interest groups that are able to offer a 'united front', or a 'universal' position, are in a substantially more advantageous position to achieve their goals from the political system. 'Internal bickering' reduces the interest group's independence from the broader political system (Baumgartner, Jones, 1993, p.176).[73] On such occasions, when a group cannot formulate a firm and effective position, and there is disagreement within the group, the importance of Level II interest representation is reduced, and replaced with an increased emphasis on Level 1 (national), and Level III (Brussels), representation. However, it is, ordinarily, only the elites that have the capability and resources to utilise all these three levels of representation.

Summary

Liberal intergovernmentalists primarily argue that political importance is a necessary prerequisite for a firm to apply successful interest representation to its own domestic government. Furthermore, informational, organisational and authoritative resources are attributes that an interest can bring to the policy formulation process. Eliteness of a firm provides such 'enabling' conditions that help facilitate the interest representation of a firm.

Grant's analysis of Euroelectric, which he accuses of being divided and slow to react to policy initiatives, leading to weak representation, can be contrasted with the substantially more successful representation of AEAC, which managed to overcome such hurdles by way of a form of qualified majority voting within its group structure. Clearly, successful interest representation requires a similar methodology of policy formulation. However, without the decision making of the interest group being weighted in the interests of the large firms, interest groups, according to Greenwood, are 'doomed to failure'. Central to Greenwood's study of the Pharmaceutical industry is that interest group effectiveness is dependent upon the large firms within the group. The Greenwood hypothesis is a crucial test in the examination of Hypothesis No. 2 of this book (air transport interest group decision making).

In summary, elite interests (in various sections of the economy as well as in air transport), are national champions and are (or once were), in state ownership. They have the resources (personnel and capital), to operate at the three levels of interest representation, depending on the circumstance. They are usually the largest firms in terms of financial turnover and employ the greatest number of employees. Primarily, they are of substantial political importance to their Member State. Non-elites have a near absolute reliance on Level II representation, and do not possess the resources to operate at Levels 1 and III when disagreement within Level II leads to a lowest common denominator of interest representation. Although some non-elites are members of trade associations, the dominance of the larger players ensures that their views are represented to a lesser extent when there is disagreement within the group. Because of the reliance by non-elites on Level II (group), representation, a dispute within the group, leading to a lowest common denominator position, can lead to a near disenfranchisement of interest representation for non-elite interests.

The theories of Moravcsik, Putnam, Huelshoff and Hix essentially state that, at the domestic level, important groups and interests influence government policy. Therefore, at the Council of Ministers (as well as other international arenas), governments strive to satisfy these domestic interests. This chapter does not dispute this prognosis. However, it may be the case that the tendency to satisfy powerful domestic interests, or an 'elite accommodation' of interests, may also exist within the European Commission. The two case studies of the formulation of a Commission Proposal for a Directive on ground-handling, and the attempted Commission Proposal on airport slot-allocation, attempt to apply the properties of liberal intergovernmentalism, a *neo-liberal intergovernmentalism,* in order to find if this situation exists within the Commission.

Notes

1 A.M. Sbragia, 'Asymmetrical Integration in the European Community - The Single European Act and Institutional Development', in D. Smith and J.L. Ray (eds), *The 1992 Project and the Future Integration in Europe* (London: M.E. Sharpe, 1993), pp.92-109.

2 Andrew Moravcsik, 'Preferences and Power in the European Community: A Liberal Intergovernmentalist Approach', in *Journal of Common Market Studies*, Vol.31, No. 4 (Dec, 1993), pp.473-524. For a critique of the Moravcsik thesis (in particular, the proposition that Moravcsik overlooks the role of the ECJ and the 'political institutions' of the EU), see: Daniel Wincott, 'Institutional Interaction and European Integration: Towards an Everyday Critique of Liberal Intergovernmentalism', in *Journal of Common Market Studies*, Vol.33, No.4 (Dec, 1995), pp.597-609. See also response: Andrew Moravcsik, 'Liberal Intergovernmentalism and Integration: A Rejoinder', in *Journal of Common Market Studies*, Vol.33, No.4 (Dec, 1995), pp.611-628.

3 Moravcsik, op.cit (1993), p.507.Examinations of the role played by domestic forces in inter-state bargaining also include: Simon Bulmer, *The Domestic Structure of EC Policy Making in West Germany* (New York: Garland, 1986); R. Rogowski, *Commerce and Coalitions - How Trade Affects Domestic Political Alliances* (Princeton: Princeton University Press, 1989).

4 Andrew Moravcsik, 'Negotiating the Single European Act: National Interests and Conventional Statecraft in the European Community', in *International Organization*, Vol.45 (Winter, 1991), pp.19-56.

5 Robert D. Putnam, 'Diplomacy and domestic politics: the logic of two level games', in *International Organization*, Vol.42, No.3 (Summer 1988), pp.427-461; See also: J. Mo, 'The Logic of Two-Level Games with Endogenous Domestic Coalitions', in *Journal of Conflict Resolution*, Vol.38, No.3 (Sept, 1994), pp.402-422.

6 Putnam's hypothesis can be applied to inter-Member State bargaining. An example is the position of France on the re-formulation of the Common Agricultural Policy. France argues at Level 1 (inter-EU Member State), that its domestic pressures restrict its 'room for manoeuvre' in the re-formulation of CAP. If Putnam's thesis is correct, the French government may well exaggerate its domestic (Level II) pressures when in negotiation, in order to achieve its desired outcome. It is, however, the case that re-formulation of CAP requires unanimity at the Council. It remains unclear if Putnam's theory would work in a scenario where policy is decided by a Qualified Majority Vote. For wide ranging discussions of the political implications of CAP, see: Secondo Tarditi, Kenneth J. Thompson, Pierpaolo Pierani and Elisabetta Croci-Angelini (eds), *Agricultural Trade Liberalization and the European Community* (Oxford: Clarendon Press, 1989). For further discussion in regard to state bargaining, see: J.L. Mortensen, 'The Interaction Between European Politics and International Bargaining: The Case of the Uruguay Round', paper presented at the European University Institute, Florence (Dec, 1995).

7 Simon Hix, 'The Study of the European Community: The Challenge to Comparative Politics', in *West European Politics*, Vol.17, No.1 (Jan, 1994), pp.1-30; p.20. See also: Barry B. Hughes, 'Delivering the Goods - The EC and the Evolution of Complex Governance', in Ray, D. Smith and J.L. Ray (eds), op.cit, pp.45-69; A. Lijphart, 'Typologies of Democratic Systems', in *Comparative Political Studies* (1968), pp.3-44; A. Lijphart, *Democracy in Plural Societies: A Comparative Exploration* (New Haven: Yale

University Press, 1977); A. Lijphart, 'Democratic Political Systems: Types, Cases, Causes and Consequences', in *Journal of Theoretical Politics,* Vol.1, No.1 (1989), pp.499-515.

8 Hix (1994), op.cit, p.1.

9 Ibid, p.20. For a discussion of the national interest, see: James N. Rosenau, Chapter 11, 'The National Interest, in James N. Rosenau, *The Scientific Study of Foreign Policy* (London: Frances Pinter, 1980), pp.283-293.

10 M.G. Huelshoff, 'Domestic Politics and Dynamic Issue Linkage: A Reformulation of Integration Theory', in *International Studies Quarterly*, Vol.38 (1994), pp.255-279. Other examinations of inter-state bargaining behaviour include: V.A. Kremenyuk (eds), *International Negotiation - Analysis, Approaches, Issues* (Oxford: Jossey-Bass, 1991); D. Druckman, 'Determinants of Compromising Behaviour in Negotiation - A Meta-Analysis', in *Journal of Conflict Resolution,* Vol.38, No.3 (Sep, 1994), pp.507-556; Peyton H.Young, *Negotiating Analysis* (Michigan: University of Michigan Press, 1991).

11 See Chapter 5.

12 Huelshoff (1994), op.cit, p.264. See also: S. Krasner, 'Approaches to the State: Alternative Conceptions and Historical Dynamics', in *Comparative Politics*, Vol.16 (1984), pp.223-224; P. Katzenstein (eds), *Between Power and Plenty* (Madison: University of Wisconsin Press, 1978); E. Nordlinger, *On the Autonomy of the Democratic State* (Massachusetts: Harvard University Press, 1981).

13 Huelshoff (1994), op.cit, p.267. See also: G. McConnell, *Private Power and American Democracy* (New York: Alfred A.Knopf, 1967); T.J. Lowi, *The End of Liberalism* (New York: W.W. Norton, 1979).

14 For a study of the relationship between German domestic politics and EC policy making, see: Stephen George, Chapter 4, 'The National Contexts: The Federal Republic of Germany', in Stephen George, *Politics and Policy in the European Community* (2nd ed) (Oxford: Oxford University Press, 1992), pp.65-81. See also: Chapter 5, 'The Party System and the Representation of Interests', in David P. Conradt *The German Polity* (5th ed) (New York: Longman, 1993), pp.83-111; Gordon Smith, Chapter 5, 'The Constraints on Politics', in Gordon Smith, *Democracy in Western Germany - Parties and Politics in the Federal Republic* (3rd ed) (New York: Holmes and Meier, 1986), pp.180-238; Emil J. Kirchner, 'The European Community: Seeds of Ambivalence', in Gordon Smith, William E. Paterson, Peter H. Merkl and Stephen Padgett (eds) *Developments in German Politics*, (London: Macmillan, 1994), pp.172-184.

15 Leon N. Lindberg, 'Financial Deregulation, Monetary Policy Co-ordination and Economic and Monetary Union', in Smith and Ray (eds) (1993), op.cit, p.117.

16 See A. Moravcsik (1993), op.cit.

17 R.M. Cyert and J.G. March, 'Organisation, Structure and Pricing Behaviour in an Oligopolistic Market', in *American Economic Review* (Mar, 1955), cited in: Grant Jordan and Andrew McLaughlin, 'The Rationality of Lobbying in Europe: Why are Euro Groups So Numerous And So Weak?' (Centre for European Studies, Discussion Paper No.7, November, 1991), p.4.

[18] Justin Greenwood, Jurgen R. Grote, Karsten Ronit, 'Introduction: the Transnational Dimension', in Justin Greenwood, Jurgen R. Grote, Karsten Ronit (eds), *Organized Interests and the European Community* (London: Sage, 1992), p.8.

[19] For detailed information, see Chapter 4.

[20] For aviation historians, the exact chronology is as follows.

Imperial Airways was formed on 31 March 1924 by the merger of four airlines; British Marine, Daimler Airways, Handley Page Air Transport and Instone Air Lines. Imperial Airways was created to develop air routes to the British Empire. Several European routes were also operated, the most important of which was London – Paris. British Airways first appeared on the scene in 1935 and operated routes from the UK into mainland Europe. In 1939 Imperial Airways and British Airways were merged by an Act of Parliament and BOAC was born. In August 1946 a separate division of BOAC was created to operate European routes; this division was called British European Airways (BEA). The 1971 Civil Aviation Act created the British Airways Board and effectively ended the separate status of the two companies, but it was not until 1 April 1974 that the 'new' British Airways was born and the company liveries of BOAC and BEA disappeared into history. For an historical background of many of the world's flag carrier airlines see: C. Chant, *Airlines of the World* (London: Tiger Books, 1997).

[21] Later re-named British Petroleum and later still privatised, BP retains a 'national champion' status in the UK. See: Geoffrey Jones, 'The 'Old Aunts': Governments, Politicians and the Oil Business', in Turner (eds) (1984), op.cit, p.161.

[22] See also: David Vogel, *Kindred Strangers - The Uneasy Relationship between Politics and Business in America* (New Jersey: Princeton University Press, 1996); Wyn Grant, 'DIY: The Government Relations Functions of Large Companies', in Grant Jordan (eds), *The Commercial Lobbyists* (Aberdeen: Aberdeen University Press), pp.99-110. For a discussion of the relationship between elected governments and the business world (in particular, the proposition that business may 'lose' from such a relationship), see: Neil J.Mitchell, *The Conspicuous Corporation - Business, Public Policy and Representative Democracy* (Michigan: University of Michigan Press, 1997), especially pp.167-189.

[23] Wyn Grant, *Business and Politics in Britain* (Basingstoke, Macmillan, 1993), p.46-47.

[24] Susan Strange argues that the 'national champion' strategy has in fact been detrimental to UK industry. Indeed she argues that the most successful UK firms are multinationals which have operated without 'government protection'. See: Susan Strange, 'Who are EU? Ambiguities in the Concept of Competitiveness', in *Journal of Common Market Studies*, Vol.36, No.1 (March 1998), pp.101-114. See also: Susan Strange, *The Retreat of the State* (Cambridge: Cambridge University Press, 1996).

[25] See: Dinos Kyrou and James Thompson, *European Domination by Incumbents? Air Transport and Telecommunications – A Reflection of Commission Decision Making* (forthcoming).

[26] This may be described as a government's 'nationalised mentality', where the government continues to treat the firm as if it was still nationalised, despite the fact that the firm has been partly, or fully privatised. This view was confirmed by a target of interest

representation within DGVII of the Commission (Transport); Interview: Jerrard Borel (Senior Administrator), DGVII, Brussels, 20.1.97.

27 Braam attempted to test the influence of small and large firms. His formula categorised firms as those with less than 100 employees, those with 100-499 and those with 500 or more employees. The study attempted to measure the 'influence attempt' by a firm and the resulting 'compliance' of the government (or success of the influence). See: Geert P.A. Braam, *Influence of Business Firms on the Government - An Investigation of the Distribution of Influence in Society* (The Hague: Mouton, 1981), p.133.

28 For a similar examination, based on the premise of the decline of the national champion, see D. Hickie, 'Airbus Industry and the Struggle in the GATT', in R.H. Pedler and M.P.C.M. Van Schendelen (eds), *Lobbying the European Union* (Aldershot: Dartmouth Publishing, 1994), pp.41-65.

29 W.E. Paterson, 'Regulatory Change and Environmental Protection in the British and German Chemical Industries', in *European Journal of Political Science*, Vol.19 (1991), pp.307-26.

30 Stephen George, 'Policy Networks' (unpublished paper, presented at the University of Aberdeen, 31.5.97).

31 This hypothesis refers to interest groups on 'consultation lists'. However, it might also be applied to interest representation *per se.* See: William A. Maloney, Grant Jordan and Andrew M. McLaughlin, 'Interest Groups and Public Policy: The Insider / Outsider Model Revisited', in *Journal of Public Policy,* Vol.14, No.1 (1994), p.25. Furthermore, there may exist a notable similarity between the traditional reliance of states on national champions for information in formulating legislation, and the European Commission's near total reliance on national champions from the Member States for information for the formulation as well as the implementation of policy. Because of the substantial lack of expertise of the various parts of the Commission on the technicalities of various policy sectors, it has been suggested that national champions will often draft Proposals that the Commission will then 'rubber stamp' (Interview: Robert Elphick (External Affairs Manager), International Diabetes Federation (based in Brussels), Aberdeen, 27.5.97). Moreover, the relatively small number of staff within the Commission increases the selectivity of the Commission for its information sources, and hence increases its dependence. Such a situation, in regards to civil aviation, is also claimed to exist in the US Congress, where the legislature is highly dependent upon air transport interests for technical advice when formulating legislation. See: Emmette S. Redford, 'A Case Analysis of Congressional Activity: Civil Aviation, 1957-58', in *Journal of Politics*, Vol.22 (1960), pp.228-258. Additionally, see Redford (1960), for an explanation as to how a relatively small staff base of a legislator increases the dependence of the legislator on external interests.

32 There exists debate as to the precise definition of a peak association. Averyt takes the term to mean an association of national interest groups. However, the term is also defined as a single group representing 'sub-groups', which are not necessarily national associations

themselves. For a discussion of the term, see: John P. Heinz, *The Hollow Core: Private Interests in National Policy Making* (Massachusetts: Harvard University Press, 1993).

33 See also, Jordan and McLaughlin (1991), op.cit, p.26.

34 For a further study of pan-EU 'policy networks' that facilitate informal bargaining and provide order to the complication of institutional policy formulation, see: H. Kassim, 'Policy Networks, Networks, and European Union Policy Making: A Sceptical View', in *West European Politics,* Vol.17, No.4 (October, 1994), pp.15-27. See also criticism of Kassim, in: J. Peterson, 'Policy Networks and European Union Policy Making: A Reply to Kassim', in *West European Politics,* Vol.18, No.2 (April, 1995), pp.389-407.

35 *Proposal* is taken to mean a Proposal of the European Commission, as opposed to the common usage of the term; Mazey and Richardson, 'Introduction: Transference of Power, Decision Rules and Rules of the Game', in S. Mazey and J. Richardson (eds), *Lobbying in the European Community* (Oxford: Oxford University Press, 1993), p.21.

36 An example of such division is that which occurred between the firms which comprised Euroelectric (an interest group representing electricity firms), specifically between its British and Continental members. This division substantially reduced Euroelectric's effectiveness in interest representation. See: W. Grant, 'Pressure Groups and the European Community', in Mazey and Richardson (1993), op.cit, p.31.

37 The costs are not simply the costs of membership of the organisation, but the costs in both direct expenditure and executive time.

38 Jordan and McLaughlin (1991) op.cit, p.49.

39 Ibid, p.48. See also: Cyert and March (1955), op.cit; H.A. Simon, 'A Behavioural Theory of Rational Choice', in *Quarterly Journal of Economics* (February, 1958).

40 See: Andrew M. McLaughlin, Grant Jordan and William A. Maloney, 'Corporate Lobbying in the European Community', in *Journal of Common Market Studies,* Vol.31, No.2 (June, 1993), pp.191-211.

41 These groups were the Liaison Committee of the Automobile Industry of the Countries of the European Communities, and the Common Market Automobile Constructors Association.

42 McLaughlin, Jordan and Maloney (1993), op.cit, p.199.

43 See: Greenwood, Grote and Ronit (1992), op.cit, pp.10-11.

44 McLaughlin, Jordan and Maloney (1993), op.cit, p.210.

45 Ibid, p.211.

46 The study explored the activities of the European Federation of Pharmaceutical Industry Associations (EFPIA), and the Senior Advisory Group Biotechnology (SAGB). W. Grant, 'Organised Interests and the European Community', paper presented at the 6th international colloquium of the Feltrinelli Foundation, Corton, 29-31 May, 1990, cited by

J. Greenwood, in 'Pharma and Biotech: Virtues and Trends in EU Lobbying', in Pedler and Van Schendelen (1994), op.cit, pp.183-198.

47 See also: Justin Greenwood and Karstin Ronit, 'Interest Groups in the European Community: Newly Emerging Dynamics and Forms', in *West European Politics*, Vol.17, No.1 (Jan, 1994), pp.31-52.

48 Averyt (1977), op.cit.

49 Greenwood, in Pedler and Van Schendelen (1994), op.cit, p.195.

50 Ibid, pp.195-196.

51 Additionally, Greenwood notes that the pharmaceutical industry of Europe consisted of a relatively high amount of *mutual interest* between the various national industries. This is in contrast to Paterson's study of the industrial chemical industry where the views of the firms differed throughout Europe.

52 EACEM was formed in 1983 as a result of the continued integration of related policy matters, in particular those relating to the standardisation of electronics. See: A. Cawson, 'Interests, Groups and Public Policy-Making: the case of the European Consumer Electronics Industry', in Greenwood, Grote, Ronit (1992), op.cit, pp.99-118.

53 Ibid. The study noted that the Chairperson of EACEM was previously a high level executive in the electronics firm Thorn EMI Ferguson for a substantial period of time. The Vice-Chairs of EACEM were from Philips and Thompson, in addition to many of its staff. The positions taken by EACEM in matters such as satellite broadcasting and High Definition Television (HDTV) reflected (not surprisingly), a high degree of similarity to the views of these particular firms.

54 See also: J. Sargent, 'Corporatism and the European Community', in W. Grant (eds), *The Political Economy of Corporatism* (London: Macmillan, 1985), p.236.

55 The telecommunications industry, similar to the airline industry, was an industry that contained a large number of national monopolies. Furthermore, it was similar in that a high degree of public control existed because of the military significance of the telecommunications industry. In most Member States the telecommunications industry was integrated into the postal service as telecommunications was seen as a commercial threat to the postal service.

56 See: Jenny Stewart, 'Corporatism, Pluralism and Political Learning: A Systems Approach', in *Journal of Public Policy,* Vol.12, No.3 (Jul-Sep, 1992), pp.243-255.

57 UNICE bulletin, No.1, 1991, cited in, Schneider, op.cit, p.67.

58 For example, a representative of an oil firm stated that the UK Offshore Oil Operators Association would offer relatively small companies a 'cheap seat at the table' to increase the legitimacy of the group (communication between representative of oil firm and G. Jordan, of the University of Aberdeen, Dec, 1995).

59 An example is the case study of the European Association for Textile Polyolefins (EATP). The study deduced that its relatively successful interest representation was derived from a

dual stranded ability to develop a reputation with the Commission for reliable data and information and the ability to supply technical expertise on the relevant standards of that field. Additionally, the interest composition of EATP was *Co-operative* giving greater legitimacy to the views of the group. See: J.P. Peckstadt, S. Mazey and J. Richardson, 'Defending and Promoting a Sectoral Interest within the European Community: The Case of the Textile Polyolefins Industry', in Mazey and Richardson (1993), op.cit, pp.177-187.

60 See Chapter 6 for details of airline alliances.

61 Travel agents are able to alternate between different airlines' Computer Reservation Systems via the same computer terminal. An airline will always show its own flights first via its own system and 80% of flights are booked from the first screen displayed on a CRS. Therefore the benefits of a flight appearing on the first screen are substantial. This Proposal stated that a code sharing 'partner', airline could not also appear on the same screen.

62 See Chapter 4.

63 British Airways fulfils the role of an elite national champion, as Chapter 4 clearly shows. Despite this however, it may be argued that British Airways' relationship with its national government, *in comparison to* the other elite airlines from across the EU, is not as intimate as these airlines, as a result of being in private rather than in state ownership. However, its relationship is considerably closer than in the case of the non-elite airlines. See Chapter 4, for further details.

64 Van den Polder, in Pedler and Van Schendelen (1993), op.cit, p.118.

65 See: references, Chapter 2, f/n1-7.

66 However, obstruction in the Council of Ministers is not an adequate substitute for influencing the content of a Proposal in its early stages. The veto at the Council of Ministers is very rarely used. In practice a 'negotiation by exhaustion' is often conducted (Grant, 1993, op.cit, 1993, p.29). There exist, therefore, severe potential disadvantages for an interest (whether elite or non-elite), in solely targeting its interest representation at its particular Member State, even in cases where Unanimity at the Council of Ministers is needed.

67 See Chapter 6, Section 4.

68 Interviews with various air transport interests which can be classified as non-elites, appear to produce support for this theory, including: Malcolm Hart (Chief Executive), Gill Airways, Newcastle, 19.3.97, and Frank Pullman (Chief Executive), Luton Airport, 6.1.97.

69 See: Neil Mitchell, 'The Decentralization of Business in Britain', in *Journal of Politics*, Vol.52 (1990), p.630. See also: Wyn Grant, *Business...* (1993), op.cit, p.87.

70 Often, such a department is classified as a 'Public Affairs' Department; for example, KLM has such a department which deals with governmental and EU representation.

71 See Chapter 4.

72 Two of these firms, Ian Greer Associates and Lowe Bell Associates are no longer under contract of BA, primarily as a result of the election of a Labour administration in April,

1997. It was felt by BA that these firms were no longer suitable in the new political 'environment' which was created at that time.

73 See also: Kenneth Meier, *Regulation: Politics, Bureaucracy, and Economics* (New York: St Martins, 1985).

2 The Changes to Treaty Provisions and the Development of DGIV

To date, the majority of interest representation literature has focused primarily upon classical notions of policy formulation. Proposals begin with an 'initiation' of policy within the Commission, which subsequently becomes a draft. The draft is then discussed between the Directorate Generals in 'inter-service negotiations' before being discussed by the specialist members of the *cabinets*. The draft proceeds to the 'special chefs' of the *cabinets* before proceeding to the College of Commissioners itself (Spence, 1997, pp.103-124). The draft, now a Commission Proposal, then follows the path of the Consultation, Co-operation, Co-decision or Assent procedure (depending upon the policy area and relevant Treaty base), via the European Parliament and the Council.[1] Such studies of the 'Brussels machinery' by policy formulation theorists and empiricists have allowed academics to establish an array of norms and 'best practices' for those interested in policy formulation from either an academic or a practical angle.[2]

Liberal intergovernmentalism and the derivatives of this theory put forward in Chapter 1 are based around the fact that it is the elite interests which ultimately determine new legislation. This Chapter suggests that the role of a particular Directorate General of the European Commission, DGIV – the Competition Directorate, may have been overlooked by contemporary interest representation literature. DGIV plays hugely prominent roles in both of the examined case studies. DGIV also has unique *quasi-judicial* powers, arguably not possessed by any other EU actor, other than the Court of Justice. Chapter 6 will test whether it is the case that liberal intergovernmentalism, or its proposed new form, *neo-liberal intergovernmentalism,* can be applied to the actions of DGIV.

Contemporary interest representation literature points to two periods of time - the signing of the Single European Act in 1986 and the Treaty on European Union in 1992 - when the number of 'lobbyists'[3] based in Brussels rapidly increased in number (Kohler-Koch, 1994, p.171).[4] However, although contemporary interest representation literature rightly points to the undoubted increase in the volume of interest representation activity, it has largely failed to pinpoint the rapidly changing role of DGIV.

Clearly, the expansion of Qualified Majority Voting in the Council upon the ratification of the Single European Act enormously strengthened the Commission's hand in its role as a Proposer of policy, thereby increasing its prominence as a target of interest representation. Furthermore, the Commission has, and will continue to establish and maintain strong relationships with interest groups and firms as they provide practical and logistical information, which is vital for the successful implementation of policy (Williamson, 1989). Subsequently, a form of 'dependency' relationship has been established between the personnel within the Commission and the groups and firms that wish to influence them (Butt Philip, 1987, p.75).

> The Commission has always been an open institution. The Commission believes this process to be fundamental to the development of the EU and its policies. This dialogue has proved valuable to both the Commission and to interested parties...Commission officials acknowledge the need for such input and welcome it.[5]

However, despite the undoubted strengthening of the Commission's hand as a result of QMV, the Commission in its role as a Proposer of policy, remains constrained by the fact that a Proposal must proceed through what may prove to be a highly reluctant Council. Because of powerful domestic pressures (and hence the application of liberal intergovernmentalism), the Council is under no obligation to transform Commission Proposals into Community law.[6]

Furthermore, the strengthening of the hand of the European Parliament upon the ratification of the Treaty of Amsterdam, with the establishment of a new form of Co-decision, or 'Simplified' Co-decision in air transport policy (in addition to other policy areas), may weaken the hand of the Commission in relation to the European Parliament. This is as a result of the greater control and input that the new Co-decision procedure will allow to the Parliament in air transport policy making.[7] Indeed the ramifications of the Treaty of Amsterdam have yet to be digested by air transport interests. Airlines continue to view the European Parliament as a 'peripheral' institution, whose importance, relative to the Commission and the Member States cannot be underestimated. Airlines have been accused by MEPs of being wholly inadequate in their representation at the European Parliament and its relevant Committee - the Transport and Tourism Committee. Conversely, *airport* interests have effectively targeted MEPs to put forward their views.[8] The establishment of 'Simplified' Co-decision for all aspects of air transport policy will have vast ramifications for those seeking to influence EC policy formulation in this area. Upon the ratification of Amsterdam, MEPs will have a greater ability to amend Proposals (if the Transport Council agrees), or ultimately to block the Proposals that it finds unacceptable.

The change in the procedure of policy formulation for air transport as a result of the Treaty of Amsterdam also presented, and continues to present a potential problem for those involved in the revision of the slot-allocation Regulation. The original 1993 Regulation proceeded via the Consultation procedure; subsequently it is intended that any revision must also proceed via the Consultation procedure.[9] However, according to Sir William Nicoll (former Director General of the Council of the European Communities in Brussels), unless the new Regulation is proposed before domestic ratification of the Treaty of Amsterdam the Regulation would need to proceed via the new Co-decision procedure. This fact creates two substantial potential problems argues Sir William. Firstly, the European Parliament is effectively 'out of action' during the first half of 1999 as MEPs and other candidates seek re/election. Following the summer break of 1999 and the MEPs becoming 'settled in' to their new positions, the European Commission is then due to be re/appointed by the year 2000. There is, therefore, considerable scope for delay in the process, in addition to the delay that has already occurred. Furthermore, if the Proposal was formulated under the Consultation procedure, the Parliament, providing it had not yet given its opinion under this procedure, could well argue that it should be able to have the power of Co-decision in the area. This is despite the fact that the Proposal would have been made under the original Consultation procedure.

The main problem is that there exist no 'hard and fast' rules on the transition between legislative procedures. The only way, according to Sir William, that potential legislative problems in this area might be avoided is if the Commission withdrew any Proposal it might formulate under Consultation and re-propose it under the Co-decision procedure, following the ratification of Amsterdam.[10] If this were to happen the European Parliament would subsequently become intrinsically linked with the revision of the Regulation in a way that has not been foreseen by air transport interests.

Regardless, this is hypothetical. It is also an example of the enormous legislative problems that can be created as a result of the Commission's inability to fulfil its legal obligations. Let us now examine the unique competencies of DGIV in order to comprehend the unique nature of this institution, and ultimately its role in the examined case studies of this book.

A 'policy making' body overlooked - DGIV

In addition to its role as a Proposer of legislation, the European Commission, as the 'Guardian of the Treaties', has the role of ensuring that EC law is correctly executed by Member States.[11] However, the Commission has traditionally been somewhat handicapped in this role. Prior to ratification of the Treaty on European Union (TEU), the Commission in a majority of policy areas, had no Treaty provisions which enabled it to exercise this 'policing role', other than the ability to make a referral to the European Court of Justice (ECJ). Furthermore, the ECJ had no ability to impose sanctions on Member States that it deemed to be guilty of violating EC law. The TEU has allowed the Commission to refer a Member State back to the ECJ if it has refused to comply with a previous ECJ judgement. The ECJ may subsequently impose a financial penalty, although the decision as to the amount and specifications of that penalty remain with the ECJ (Nugent, 1992, pp.114-115).[12] However, this financial sanction available to the ECJ, has never been applied, although the Commission has referred Member States to the ECJ on a multitude of occasions.[13]

23 of the 24 composite DGs of the European Commission remain largely *legislative*; their task being to formulate draft Proposals to put before the College of Commissioners. Subsequently, there is the possibility that the 23 legislative elements of the Commission (such as DGVII, Transport), face a future where their 'room for manoeuvre' in Proposal formulation is not only *de facto* limited by the political constraints of the Council, but arguably also by the political constraints of what might be, in some policy areas, a hostile European Parliament.

Conversely, DGIV is unique amongst the Directorates. Its primary role is to ensure, by way of a Commission Decision or a Commission

Directive, that the primacy of existing legislation, namely the competition provisions of the EC Treaty, are executed and not abused by firms, or indeed by Member States. DGIV has the ability to impose sanctions on Member States or firms by way of a Commission Decision, without firstly referring the issue to the European Parliament, Council or European Court of Justice[14] (although all Commission Decisions are open to appeal at the ECJ).[15] Subsequently, DGIV is primarily *quasi-judicial,* as opposed to legislative, in its nature.[16] That said, all Commission Decisions and Directives require a simple majority of the College of Commissioners in order to become legally binding.[17] Therefore DGIV does not operate in a political or indeed an operational vacuum, and the Competition Commissioner relies on the support of his colleagues in order to pursue the interests of DGIV. However, if such support is not forthcoming there are other ways and means by which DGIV can pursue its intended policy. Two of these particular ways are examined below, but firstly there needs to be a closer examination of the primary role of DGIV in order to comprehend its role in the case studies of the formulation of the Commission Proposal for the ground-handling Directive, and attempted revision of the slot-allocation Regulation.

The role of DGIV

Each Member State of the European Union possesses a competition authority to enforce national competition laws. However, DGIV 'polices' the single market, 'to ensure that its benefits reach the ordinary citizen'.[18] The roles of DGIV, and its exclusive powers, are central to the study of the application of the hypotheses. It is therefore necessary to examine more closely these exclusive powers, before progressing to the case studies.

DGIV has four main areas of responsibility.

1. The prevention of cartels.
2. The regulation of state aids.
3. To vet, or block, mergers and take-overs.
4. To prevent abuse of a dominant market position.[19]

1. Cartels, where prices are fixed, either formally, or 'informally' (such as in a 'gentleman's' agreement') are outlawed by Article 85 of the EC Treaty if they distort competition between Member States.

2. A 'state aid' or public subsidy to a firm could potentially threaten competition between Member States. If so, such a subsidy would come into conflict with Articles 92-94 of the EC Treaty. The Commission has the ability to vet, and ultimately prevent all state aids, if it believes that such an aid would distort competition. However, in practice, the majority of state aids are 'permissible' as in the view of the Commission they are aimed at 'developing the interests of the Union as a whole'. Such permissible aids are normally linked with regional development objectives, or targeted developments in a particular section of industry.[20]

3. There is no provision specifically within the EC Treaty that gives the Commission the competence to vet mergers and take-overs. However, mergers between companies can create monopolistic market structures, by reducing the level of competition that exists in a particular segment of the market. Therefore, there has been widespread support for the theory since approximately 1970, that Article 85 (prevention of cartels), and Article 86 (prevention of the abuse of a dominant market position), could legally be applied to merger situations.[21] However, it was not until the catharsis provided by the Single European Act that the Council acted on a Proposal by the Commission, and adopted the 1989 Council Merger Regulation.[22] The Regulation applies to all industries in mergers with a 'Community Dimension'. Two conditions must be met to fulfil this criterion.

1. The aggregate world-wide turnover of the firms concerned is over 5 billion Euro, or,

2. The Community-wide turnover is over 250 million Euro.

At the time of writing, eight mergers have been blocked absolutely.[23] The vast majority of mergers are however permitted subject to modification as a result of a DGIV examination.[24]

According to many commentators, airline alliances[25] are *de facto* mergers.[26] Two or more airlines in an alliance sell seats on each others' flights, at an identical price, and therefore price competition between the airlines is eliminated. Since the 'three packages' of airline de-regulation,[27] the European Commission has had the competence to vet and ultimately reject alliances that take place *within* the EU. Until 1996 the European Commission did not vet alliances between airlines based in EU Member States and airlines based in third countries, such as the US. The competition implications of such agreements were decided on a bilateral basis between the Member State and the third country. However, in June 1996 British Airways announced a planned alliance between itself and the carrier American Airlines; this proposed alliance was to have substantial implications for the attempted Proposal for the slot-allocation Regulation. The European Commission announced an investigation into this alliance together with five other Member State - US alliances that already existed.[28] The legal basis for this announcement, argued DGIV, was Article 89 of the EC Treaty.

Article 89 stipulates that the Commission must ensure the principles of competition, as laid out in Articles 85 and 86, are carried out in the Member States. The Commission, in co-operation with the relevant Member State authorities shall investigate suspected cases of infringement of these principles. If an infringement has taken place the Commission shall propose appropriate measures to bring the infringement to an end. Crucially, the Commission may formulate a Decision on the matter if the authorities within the Member State concerned do not take appropriate action. Article 89 is therefore extremely broad in its interpretation, and essentially states that if the Commission does not deem Member State action appropriate, the Commission can subsequently take further action. Article 89 was used in the British Airways / American Airlines situation.[29]

4. 'Market dominance' was a key ingredient of the involvement of DGIV in the two case studies under examination. Holding a dominant market position does not violate the competition provisions of the EC Treaty; however abuse of this dominance is such a violation. A dominant undertaking must be proved to be dominant in a narrowly specified market. Abuse of position is judged to have taken place if the dominant undertaking has, as a result of its behaviour, weakened the degree of competition.

Definition of dominance

The European Court of Justice has not set a specific percentage of the possession of a market at which an undertaking becomes dominant. The Court has however stated that dominance occurs when a firm enjoys, 'a...position of economic strength...which enables it to prevent effective competition being maintained in the relevant market...'.[30]

Soames states:

> An undertaking will be found to be dominant if it enjoys a position of economic strength which enables it to hinder the maintenance of effective competition on the relevant market by allowing it to behave to an appreciable extent independently of its competitors, customers and ultimately consumers. A market share of over 50% may be regarded as giving an undertaking a dominant position.[31]

Woods argues that 75% of a market share provides strong evidence of dominance, although there exists evidence in some markets that an undertaking with 40% of the market share is in a dominant position.[32] This substantial variation depends largely on the individual market share of the undertaking's competitors, as well as other factors, which may often be unique to a particular type of market.[33] The Commission, on an extremely precise and narrow basis, assesses dominance in a particular market. 'Demand substitutability', is the key test by which the Commission identifies a particular market; i.e. the choice that a consumer (for example an airline passenger, or, for that matter, an airline, which is a consumer of ground-handling services), has in a narrowly defined market place.

In *Ahmed Saeed* the Court stated:

> The test to be employed is whether the scheduled flight on a particular route can be distinguished from the possible alternatives by virtue of *specific characteristics.*[34]

The case of *British Midland v. Aer Lingus* concerned the actions of Aer Lingus in attempting to end the ground-handling agreement that it held with British Midland at Heathrow Airport. This would have resulted in British Midland being unable to operate out of Heathrow, and having to switch to another London airport. Aer Lingus argued that this did not constitute abuse of a dominant market position (refusal to supply; see below), as Aer Lingus was not a dominant ground-handler at London's airports. However, the Commission considered the demands of a particular class of customers, namely, business passengers, when determining the relevant market. Although the Commission recognised the substitutability of London airports in respect of London - Dublin flights, it found that, in the market of business passengers on the London - Dublin route, such passengers,

> ...traditionally prefer Heathrow, amongst other reasons because other London airports are not served as frequently as Heathrow, and because they do not offer a similar range of connections and therefore will be less suited to onward travel beyond London.[35]

Therefore the Court established that for dominance in a market to be proved the complainant must very specifically identify the particular market.

Abuse of a dominant position

If dominance is established, the undertaking will be considered to be abusive if, as a result of its behaviour, competition in the market is weakened. Examples of abusive behaviour include predatory pricing, unfairly high pricing and refusal to grant access to an essential facility. These examples are examined in turn below.[36]

Predatory pricing

Predatory pricing is when an undertaking (such as an airport) either sets out to reduce another firm's profitability in order that the firm has no choice but to exit the relevant market, or sets out to prevent another firm entering the market in the first place. Predatory pricing can be proved in one of two ways.

1. If evidence exists (such as a company memo) that the aim of an undertaking was to deliberately achieve the above then the Commission may take appropriate action.

2. If no such documentary evidence exists, the European Court of Justice has tested what constitutes abuse in the case *AKZO*.[37]

In this case the Court stated that if a dominant undertaking sets its prices at below the Average Variable Cost,[38] the undertaking would be in abuse of Article 86, and the Commission may subsequently take appropriate action. In practice, the *AKZO* test is extremely difficult to prove, as there are a substantial array of financial accounting formulae that make this task extremely difficult.[39]

Unfairly high pricing

Unfairly high pricing is where an undertaking, such as an airport, sets prices to achieve turnover and profits which, in a more competitive environment it would be unable to do. The test case, argues Soames, is *United Brands*.[40]

The Court asked:

> ...whether the difference between the costs incurred and the price charged is excessive, and [if so] to consider whether a price has been charged which is either unfair in itself, or as compared to the other competing products.[41]

Refusal to supply access

Refusal to supply access is a particularly useful example for the case study of ground-handling. If an undertaking both owns, and itself uses an essential facility (such as an airport terminal), and refuses to grant access to that facility to a competitor, or will grant access to the competitor only subject to less favourable terms and conditions, this would place competitors at a disadvantage and may infringe Article 86.

The Commission has defined an essential facility as, '...a facility or infrastructure without access to which competitors cannot provide services to their customers.'[42]

A number of test cases exist which have established this principle. In *Holyhead I*[43] the operational practices of the Port of Holyhead were changed to the advantage of the owners, Stena Sealink. The ferry company B&I, a user of the port, argued that the new operation of the port would discriminate against its services. The Commission concluded that the port was an essential facility for B&I, as there were no other ports that it could reasonably substitute Holyhead for B&I to use as an alternative. In *Holyhead II*[44] the firm Sea Containers, alleged, amongst other things, that Stena Sealink had refused Sea Containers access to the port for the operation of its 'SeaCat' service. The Commission found against Stena Sealink in both *Holyhead I* and *Holyhead II.* Similar cases relating to ports are *Roscoff,*[45] where the Commission adopted interim measures against the owners, the Morlaix Chamber of Commerce, by finding that it had abused its dominant position by refusing Irish Ferries access to port facilities (contrary to Article 86), and *Elsinore,*[46] where the Danish shipping line Mercandia was refused permission to operate a car ferry service by the port's owners. Soames notes that the Commission has shown itself willing to take action where it suspects that the actions of the essential facility operators were not motivated by 'legitimate' operational concerns, but by a desire to advantage its operation in the relevant market.[47] Port access has found itself to be the main testing ground in cases of refusal to grant access, and a substantial body of *jurisprudence* has been built up in this area.

The above cases have an important relevance in both the case studies of slot-allocation and ground-handling. In slot-allocation, if the airport operator is connected in any way to the flag carrier airline and allocates the most favourable slots to that airline, or charges discriminatory fees, this would be considered an abuse.[48] In the case study of ground-handling, the Commission was confident that the body of *jurisprudence* built up on port access was sufficient to warrant a Commission Directive based upon Article 90 of the EC Treaty.

Article 90

Article 90 states that, where the Member State owns an undertaking, or the Member State has granted special or exclusive rights to the undertaking, the Member State shall not,

> ...enact or maintain in force any measure contrary to the rules of the Treaty, in particular...Articles 85 to 94 [namely the competition rules].[49]

Some industries, the Commission recognises, must be publicly funded, supported and protected from the rigours of competition. For example, ferry companies may operate routes to peripheral areas with a public subsidy as the routes would otherwise be non-viable. In a similar way, Member States may award Public Service Obligations for air services (subject to various rules on tendering), on routes which not be commercially viable.[50] The Commission also allows postal services to be offered universally and at an affordable rate throughout the territory of the EU, which includes rural or remote areas which would also be non-viable to operate on a commercial basis.[51] Such activities, *do not* constitute a violation of a dominant market position, as Article 90 (2) provides the exemption that,

> ...a monopoly shall be subject to the rules contained in this Treaty, in particular, the rules on competition, in so far as the application of such rules does not obstruct the performance, in law or in fact, of the particular tasks assigned to them.[52]

However over time, due primarily to technological developments and changes in the structure of markets, industries that have traditionally needed the protection of being a governmentally maintained monopoly may no longer be able to argue that they require such protection. An example of this situation is the progression in the telecommunications industry. Traditionally, states protected their telecommunications markets in order to cover the non-profitable nature of maintaining and operating a telecommunications network.[53] Telecommunications were protected from the rigours of free market economics as the industry was deemed a public service. However, the technological developments in telecommunications have meant that the exemptions from competition granted to Member States by Article 90 (2) were no longer valid, as telecommunications became an industry which could operate in a commercial environment without the need for protection from the market. Therefore, the Commission adopted a Directive in 1990 which liberalised the telecommunications sector by clearly and categorically bringing telecommunications out of any possible exemptions under Article 90 (2), and bringing them firmly under the competition auspices of the EC Treaty.[54] It is Article 90 (3) that grants the Commission the competence to formulate Directives or Decisions in this area.

Article 90 (3) states:

> The Commission shall ensure the application of the provisions of this Article and shall, where necessary, address appropriate Directives or Decisions to Member States.[55]

In the case of ground-handling liberalisation, the Commission had the opportunity to formulate a Directive to liberalise the ground-handling market in the way that it had liberalised the telecommunications market. However, the Commission did not take advantage of this opportunity; the reasons for this are examined in Chapter 6.

Procedures of DGIV

1. *Complaints*

Any interested person, either natural or legal (for example, a firm), may complain about anti-competitive behaviour to DGIV. Such complaints were made by airlines on the issue of ground-handling.[56] Interested persons may also approach the national courts, as the competition rules are directly applicable to all Member States unless the Commission has already initiated proceedings, in which case the national courts must wait for any Commission ruling (or ECJ ruling). If the Commission decides not to officially act then a national court must interpret the relevant section of the Treaty of Rome in the way that it sees fit.

2. *Own initiative*[57]

DGIV may, upon its own initiative, investigate issues of competition. The initiative may have been sparked by information from national authorities, interest groups, by the press coverage of a particular issue, or even by an individual within the firm under question. Such 'Commission initiative' was used in the investigation of the proposed alliance between British Airways and American Airlines (of direct relevance to the revision of the slot-allocation Regulation), prior to any complaints being made to DGIV on the issue.[58]

If DGIV decides to commence an investigation into a particular firm it will request information from relevant parties. Relevant parties that fail to provide adequate information upon request will receive a formal request by the Commission. Continued failure can result in fines of up to 1000 Euro per day.[59] The Commission has the power to send officials who can examine records, take copies, and ask for oral 'on the spot' explanations. DGIV officials may enter any premises, land or vehicles belonging to the subject of the investigation. In certain circumstances the Commission may launch a 'dawn raid' where there is a sudden, unannounced visit to the company, and documents and computer records may be seized. In particular, such action may be taken when the Commission believes that information may be destroyed.[60] The Commission has the ability to fine firms in breach of the competition rules up to 10% of the undertaking's global turnover in the previous financial year. In determining the fine, the Commission will take into account the seriousness of the breach, the length of time that it continued, and the size and market share of the firm concerned.[61]

The limitations and potential opportunities of DGIV

Although Member States are technically separated from the competition actions of the Commission, it is highly unlikely that Member States and their composite elite interests would not be involved, at least to a certain degree with the actions of the Commission. Indeed, the study of the Proposal for the ground-handling Directive and the attempted Proposal on slot-allocation test the extent of this Commission 'independence' from the Member States. However, despite the fact that the Commission does not formulate Decisions and Directives in a 'political vacuum', DGIV has made effective use of other avenues to pursue its goals. Two of these avenues are the use of Judicial Review, and the use of 'Secret Jurisprudence', both of which are examined below. Furthermore, both bear a direct relationship to the options available to DGIV in the examined case studies.

1. *Court Rulings (Judicial Review)*[62]

Court rulings create a body of jurisprudence that enables DGIV to enforce its will. Wilks argues that the time taken to build-up a body of jurisprudence is a contributory factor to the relatively 'late development' of DGIV as an institution.[63]

Hoffman-La Roche 1979,[64] provided the jurisprudence which today enables DGIV to apply a definition of a dominant market position. Similarly, *AKZO* 1991,[65] set the jurisprudence which determines the calculation by which an undertaking has abused its dominant position by way of predatory pricing.

In the air transport sector, *Nouvelles Frontières* 1986[66] and *Ahmed Saeed* 1989[67] created jurisprudence which allowed DGIV the 'mandate' to formulate a Commission Directive to liberalise the air transport sector. Although the Commission did not formulate such a Directive, these cases were a highly significant contributory factor which 'forced' the Member States to adopt the 'three packages' of air transport liberalisation by way of Council Regulations. The cases placed the Commission in a position where it could enforce air transport competition *without* the need for agreement from a highly reluctant Council of Ministers who, in the main, wished to protect their own nationally based flag carriers. The Member States were therefore in a position where the Commission would formulate its own rules on air transport unless the Council formulated rules by way of Council Regulations.[68]

Therefore, Court decisions allow DGIV a competence in a particular policy sector by way of jurisprudence. *Nouvelles Frontières* and *Ahmed Saeed* provided the jurisprudence needed for DGIV to ensure that any form of fare setting for flights within or originating in the European Union, was deemed illegal.

2. *'Secret' Jurisprudence*

As stated above, all Commission Decisions and Directives, subsequently published in the Official Journal, require a simple majority of the College of Commissioners. The Commission may also issue Notices that give an indication of the Commission's view on certain matters, and are designed to provide guidance. However, Commission Notices are not legally binding. Notices do not require publication in the Official Journal, although some are published in order to provide guidance.[69] DGIV may also issue informal letters to undertakings known as 'Comfort Letters'. Such letters have no legal standing, and are not published in the Official Journal. Yet the vast majority of cases that DGIV views as potential violations of competition are dealt with this way. Van Bael estimates that from 1975 to

1984, 96% of cases were 'settled' in this way.[70] Officials within DGIV informally advised undertakings that their actions were a violation of competition law, and remedies were subsequently taken to the satisfaction of DGIV.[71] Wilks estimates that for the period 1985 -1989, this figure had risen to 97%,[72] and since 1989, the proportion of cases dealt with in the informal way of 'Comfort Letter' has risen,

> ...considerably beyond the 97% figure. The vast majority of DGIV's actions take place behind the scenes. One only sees the tiny tip of a huge iceberg.[73]

Van Bael states that this situation results in DGIV being 'prosecutor, judge and jury', and leads to the creation of a '...secret jurisprudence', within DGIV.[74] Wilks argues that this situation, where officials have the discretionary powers to negotiate a settlement, occurs outside the due process contained in the 1962 Council Regulation 17[75] which laid out the procedures under which the competition auspices of the Commission were to be executed.

In air transport, since the onset of the three packages, there have been numerous occasions where complaints have been made to DGIV, Comfort Letters issued, and remedying action taken. Two examples of this are complaints made by the airlines Easyjet and Virgin Express, both in 1996. In the former case Easyjet complained to DGIV about alleged practices of abuse of a dominant position involving the airline KLM at Schiphol Airport Amsterdam. In the latter case Virgin Express complained to DGIV alleging abuse of a dominant position by predatory pricing, concerning the alleged pricing policy of the airline SAS on the Copenhagen - Brussels route. In the former case, DGIV launched a 'dawn-raid' on the premises of KLM as a result of the seriousness of the alleged complaint, which subsequently received much media coverage.[76] In the latter, far less public episode, SAS were, allegedly, informally given a warning, issued by Comfort Letter, concerning its conduct. However, in both cases, the episodes were resolved to the satisfaction of the complainants without any 'official' Decision needing to be formulated by the Commission.

Summary and discussion of the powers of DGIV

Docksey and Williams refer to *comitology* as

> ...an institutional compromise between the need for greater Commission decision-making and Member States' desire to preserve national influence over Commission decision-making.[77]

It is clear that the variation of comitology applied to DGIV by the Member States is substantially different to that applied to the other 23 Directorate Generals of the European Commission. DGIV may, by use of Directives and Decisions, enforce the competition provisions of the Treaty of Rome. However the prominence of DGIV has been built up, particularly in the past two decades by the use of jurisprudence (utilising the European Court of Justice), and by 'secret jurisprudence' by way of unofficial 'Comfort Letters.' The *unofficial* powers of DGIV substantially exceed its *official* powers, which are, in themselves, substantial. If it is the case, as Wilks argues, that the use of Comfort Letters and secret jurisprudence exceeds the competence that the Member States wished to grant DGIV by way of Regulation 17, then the definition of comitology understood by DGIV and the definition understood by the Member States may be substantially different to one another.

Traditional interest representation analysis of both an empirical and a theoretical nature has emphasised the importance and dominance of large firms (elites), in interest representation and policy formulation. However, the role of DGIV presents an under-examined counter argument to such theories. DGIV uses its substantial degree of autonomy from the Member States, and the precedents set by jurisprudence and by Comfort Letter / 'secret jurisprudence' to achieve its goals. Contemporary interest representation theory is based upon the traditional notion of 'Commission proposes, Council disposes'. Resultantly, the *quasi-judicial* nature of DGIV is largely ignored. Liberal intergovernmentalism concentrates on the formulation of Council Regulations and Directives; subsequently it concentrates on the interaction between interests and the Member States that comprise the Council. The Member States, being beholden to the dominant interests within them determine the eventual outcome of this policy. However, liberal intergovernmentalism may ignore the *quasi-judicial* nature of DGIV. It may also largely ignore the *judicial* nature of the European Court of Justice, and the primacy of Treaty obligations over Council Directives and Regulations.

Notes

1 Discussions of the European Parliament and its procedures prior to the Treaty of Amsterdam can be found in: Richard Corbett, Francis Jacobs and Michael Shackleton, *The European Parliament* (3rd ed) (London: Catermill, 1995), esp p.197; Claudia Hubschmid and Peter Moser, 'The Co-operation procedure in the EU: Why was the European Parliament influential in the decision on car emissions standards?', in *Journal of Common Market Studies,* Vol.335, No.2 (June 1997), pp.225-242; Martin Westlake, *The Parliament and the Commission: Partners and Rivals in the European Policy-Making Process* (London: Butterworths, 1994); Martin Westlake, *A Modern Guide to the European Parliament* (London: Pinter, 1994); Martin Westlake, 'The Commission and the Parliament', in Edwards and Spence (eds) (1997), op.cit, pp.239-263; S. Williams, 'Sovereignty and Accountability in the European Community', in R.O. Keohane and S. Hoffmann (eds), *The New European Community* (Oxford: Westview Press, 1991), p.162; S. Mazey and J. Richardson, 'Introduction: Transference of Power, Decision Rules and Rules of the Game', in S. Mazey and J. Richardson (eds), *Lobbying in the European Community* (Oxford: Oxford University Press, 1993), p.9.

2 See in particular: Svein S. Andersen and Kjell A. Eliassen (eds), *Making Policy in Europe* (London: Sage, 1993); S. Mazey and J. Richardson (eds) (1993), op.cit; S. Mazey and Jeremy Richardson, 'The Commission and the lobby', in Edwards and Spence (eds) (1997) op.cit, pp.178-198; Justin Greenwood, Jurgen R. Grote and Karsten Ronit, 'Introduction: the Transnational Dimension', in Justin Greenwood, Jurgen R. Grote and Karsten Ronit (eds), *European Casebook on Business Alliances* (London: Prentice Hall, 1995); Juliet Lodge, 'EC policy making: institutional dynamics', in Juliet Lodge (eds), *The European Community and the Challenge of the Future* (2nd Ed) (London: Pinter, 1995), esp. pp.28-33; H. Wallace and W. Wallace, *Policy-Making in the European Union* (Oxford: Oxford University Press, 1996).

3 An array of criteria has been used in attempting to define the terms 'lobbyist' and 'interest'. In Washington DC, where interest representation is an established norm, the definition of a lobbyist, as applied by the US 1945 Lobbying Act, has proved impossible to uphold (see: Justin Greenwood and Andrew M. McLaughlin, 'The Management of Interest Representation in the European Union', in *Journal of Common Market Studies,* Vol.33, No.1 (Mar, 1995), pp.143-156). Likewise, the European Commission has not produced an exact definition of the terms lobbyist or interest representation. Consequently, the formulation of legislation regarding the activity is extremely complex, and subsequently very little legislation had been formulated to regulate such activities. The European Parliament has however defined the term lobbyist as: '...anybody who acts on the instructions of a third party and sets out to defend the interests of that third party to the European Parliament and other Community institutions' (Greenwood and McLaughlin (March, 1995), op.cit, p.144). Following a report by UK Labour MEP Glyn Ford on the conduct and procedures of MEPs and their associations with interests, the EP drew up a specific 'code of conduct' for interests in 1997. Primarily, any interest which seeks to make representations for more than five days must register themselves and make clear

their associations (see: 'Provisions governing the application of Rule 9 (2), Lobbying Parliament', O.J. L 49, 19.2.97. See also: Justin Greenwood, 'The Regulation of Interest Representation' (specifically with reference to the Ford Report), in Justin Greenwood, *Representing Interests in the European Union* (Basingstoke, Macmillan, 1997), pp.90-100, and Sir William Nicoll, 'The Code of Conduct of the Commission Towards the European Parliament', in *Journal of Common Market Studies*, Vol.34, No2, Jun, 1996, pp.275-282).

[4] Although it is clear that there has been a substantial increase in interest representation activity, the precise number of 'lobbyists' remains far from clear. Andersen and Eliassen (1993, op.cit, p.39), estimate that the number currently exceeds 10,000. However, David Martin MEP has stated, 'This figure of 10,000 [lobbyists] has just been plucked out of the air. The figure is probably closer to 3000, with 30 or 40 major lobbying firms.' (David Martin MEP, BBC Radio Scotland, 11.6.96).

[5] J. Greenwood, 'Pharma and Biotech: Virtues and Trends in EU Lobbying', in R.H. Pedler and M.P.C.M. Van Schendelen (eds), *Lobbying the European Union* (Aldershot: Dartmouth Publishing, 1994), p.184.

[6] This book concentrates on the dynamics of Commission Proposal formulation. However, for a thorough examination of the Council's policy making provisions, post ratification of the 1997 Draft Treaty of Amsterdam, see: Andrew Duff, Chapter 2, 'The EU institutions and the budget', in Andrew Duff (eds), *The Treaty of Amsterdam* (London: Sweet and Maxwell, 1997), pp.36-69, especially, pp.41-45.
For pre-Amsterdam examinations of the Council, see the following: William Nicoll and Trevor C. Salmon, *Understanding the New European Community* (Hemel Hempstead: Harvester Wheatsheaf, 1994), pp.55-64; Fiona Hayes-Renshaw and Helen Wallace, *The Council of Ministers* (Basingstoke: Macmillan, 1997); Martin Westlake, *The Council of the European Union* (London: Catermill, 1995); M.P.C.M. Van Schendelen, ''The Council Decides': Does the Council Decide?', in *Journal of Common Market Studies*, Vol.34, No.4 (Dec, 1996), pp.531-548; Madeleine O.Hosli, 'Coalitions and Power: Effects of QMV in the Council of the European Union', in *Journal of Common Market Studies*, Vol.34, No.2 (Jun, 1996), pp.255-273.

[7] See: Article 84 (Sea and Air Transport), General Secretariat, Council of the European Union: 'Amsterdam European Council, Draft Treaty', 1997 , pp.117-119. A discussion on the ramifications of increased powers of Co-decision for the European Parliament can be found in: Andrew Duff (eds), *The Treaty of Amsterdam* (London: Sweet and Maxwell, 1997), esp. pp.149-150.

[8] Interview: Barry Seal MEP (*Rapporteur*, European Parliament Transport and Tourism Committee), Brussels, 6.1.98; Brian Simpson MEP (Member of the Transport and Tourism Committee), Brussels, 20.4.98.

[9] 'Council Regulation 95/93 on common rules for the allocation of slots at Community airports', O.J. L 14/1, 12.1.93.

[10] Interview: Sir William Nicoll (former Director General of the Council of the European Communities) (via telephone), London, 8.6.98.

[11] Article 155/169 EC Treaty.

[12] See TEU, Articles 155-163 (on the Commission), and Articles 169-171 (on the ECJ).

[13] However, as Dinan argues:

> Member States generally respect the Treaties; otherwise the Community would collapse. For political reasons, the Member States and the Commission are generally reluctant to pursue cases all the way to the Court of Justice, and most disputes are resolved at an early stage.

Conversely, Dinan also points out that deliberate non compliance exists, notably in the area of competition policy. See: 'The European Commission', in Desmond Dinan, *Ever Closer Union?* (London: Macmillan, 1994), p.225.

[14] It has this ability because of the unique competencies awarded to it by Council Regulation 17 of 1962. See: Articles 1-25, Council Regulation No.17, 6.2.62, 'Regulation determining the procedures by which DGIV can execute Articles 85 and 86 of the Treaty'. O.J. 13, 21.2.62, p.204/62. See also: J. Temple Lang, 'Community Antitrust Law - Compliance and Enforcement', in *Common Market Law Review,* Vol.18, No.3 (Aug, 1981), pp.335-262. For a wide ranging discussion of the perceived advantages, disadvantages and possible modifications of DGIV see: Alan J. Riley, 'The European Cartel Office: A Guardian Without Weapons', in, *European Competition Law Review,* Vol.18, No1 (Jan/Feb, 1997), pp.3-16.

[15] See Article 17, Regulation 17, op.cit.

[16] Although DGIV does initiate Proposals to go before the Council of Minters. Council Regulation 17, which sets out the procedures by which DGIV operates, and Council Regulation 4064/89, which sets out the procedures by which DGIV can control mergers were both initiated by DGIV.

[17] Although, as examined below, there are other methods by which DGIV can enforce its will, without requiring any use of the College.

[18] DGIV information, on: (web site) http://europa.eu.int/pol/comp/en/info.htm
See also: Barry J. Rodger and Stuart R. Wylie, 'Taking the Community Interest Line: Decentralisation and Subsidiarity in Competition Law', in *European Competition Law Review,* Vol.18, No.8 (Nov, 1997), pp.485-491; Simon Bishop, 'The Modernisation of DGIV', in *European Competition Law Review,* Vol.18, No.8 (Nov, 1997), pp.481-484; Laraine Laudati, 'The European Commission as regulator: the uncertain pursuit of the competitive market', in Giandomenico Majone, *Regulating Europe* (London, Routledge, 1996), pp.229-261.

[19] Stephen Wilks, 'Models of European Administration', Conference of the European Group of Public Administration, 2-5.9.92, pp.3-4.

[20] See: Simon Bishop, 'The European Commission's Policy Towards State Aid - A Role for Rigorous Competitive Analysis', in *European Competition Law Review,* Vol.18, No.2 (Mar, 1997), pp.84-88.
See also: (web site) http://europa.eu.int/pol/comp/en/info.htm

[21] Donncadh Woods (DGIV), 'An Outline of Community Competition Policy', paper presented to Irish Institute for European Affairs, University College Dublin, 22.3.96, on (web site) http://europa.eu.int/en/comm/dg04/speech/six/en/sp96008.htm

[22] 'Council Regulation (EEC) No.4064/89 of 21 December 1989 on the control of concentrations between undertakings', O.J. L 395, 21.12.89, p.1.

[23] Since 1990 only 8 of the 770 merger decisions have resulted in a merger being absolutely blocked (figure accurate on 28.2.98). Source: DGIV information: (website) http://europa.eu.int/en/comm/dg04/merger/closed/en/mergstat.htm
Arguably the most notorious decision (because the view of the Commission ruled against the express wishes of the French and Italian governments), was that to block the proposed merger in 1990 between the aerospace companies, Aérospatiale/Alenia/De Havilland (Ref. IV/M.052).

[24] D. Woods (DGIV), op.cit, pp.5-6.

[25] For a detailed analysis of existing airline alliances see: 'Airline alliances', in *Airline Business* (June 1998), pp.40-48.

[26] See: Philippe Chappatte (Partner, Slaughter and May), 'The Alliance Between SAS and Lufthansa - A Model For Future Co-operation Between Airlines in Europe', paper presented at the European Air Law Association, Copenhagen, 8.11.96.

[27] See Chapter 3.

[28] See Chapter 6, Section 4.

[29] The reasons why the Commission used Article 89 in this situation are examined in detail in Chapter 6. However, an analysis of the legal nature of the Commission's use of Article 89, with reference to the proposed alliance between British Airways and American Airlines, can be found in: Christopher Kerse, 'Enforcing Community Competition Policy Under Articles 88 and 89 of the EC Treaty', in *European Competition Law Review*, Vol.18, No.1 (Feb, 1997), pp.17-23. See also: Rodger and Wylie (1997), op.cit.

[30] *Hoffmann-La Roche*, ECJ 13.2.79, ECR 461, 520.

[31] Trevor Soames (Partner, Norton Rose), 'Access to airport facilities, the impact of competition law', paper presented at the Royal Aeronautical Society, 'Slots and Access to Airport Facilities', conference, London, 21.1.97, p.6-7.

[32] However, although there is no actual figure confirming the percentage of a market that constitutes dominance, the Court of Justice has stated that 10% would *not* constitute dominance; see: *Metro v. Commission*, Case 26/76R [1976].

[33] Donncadh Woods (DGIV), op.cit. It should be noted that the initial complaints to the Commission concerning ground-handling were aimed at a situation where the undertaking was alleged to enjoy a 100% market share.

[34] Case 66/86 *Ahmed Saeed* [1989], ECR 803 [1990] *Common Market Law Review*, Vol.4, No.102.
For dominance identification see also: Case C-333/94 P, *Tetra Pak International SA v. Commission*, 1996.

[35] *British Midland v. Aer Lingus* [1992] O.J. L 96/34 [1993] *Common Market Law Review*, Vol.4. No.516. See also: Bernardine Adkins (Dechert, Price and Rhoads), 'Use of Article 86 to Secure Market Access', paper presented at IBC, European Air Transport Services and Airports, conference, Brussels, 6.2.97, p.4.

[36] Trevor Soames (Partner, Norton Rose), 'Access to Airport Facilities - The Impact of Competition Law', paper presented at the Royal Aeronautical Society, 'Slots and Access to Airport Facilities', conference, London, 21.1.97, p.7.

[37] *AKZO,* Case C-62/86 [1991] E.C.R. I-3359.

[38] The Average Total Cost is a firm's fixed expenses (for example, rent), plus variable expenses (the extra cost of producing or servicing each extra unit). The Average Variable cost is simply the cost to a firm of producing or servicing each extra unit.

[39] Soames, 'Access...', op.cit, 1997, p.7.

[40] *United Brands,* Case C-27/76 [1978] E.C.R. 207.

[41] Soames, 'Access...', op.cit, 1997, p.7.

[42] *B&I Line plc v. Sealink Harbours Limited and Sealing Stena Limited* [1992], *Common Market Law Review,* Vol.5, No.255., 1992.

[43] Ibid.

[44] *Sea Containers v. Stena Sealink - Interim Measures:* O.J. L 15/8 [1994].

[45] European Commission: XXV Report on Competition Policy, p.120.

[46] Commission Press Release IP 96/205.

[47] Soames, 'Access...', op.cit, 1997, p.11.

[48] See: Article 8 ('Process of slot-allocation'), 'Council Regulation (EEC) No.95/93 on common rules for the allocation of slots at Community airport'. O.J. L 14 22.1.93.

[49] Article 90.1, Treaty of Rome.

[50] For further details see: Dinos Kyrou, 'Developments in Public Service Obligations', paper presented at the IBC Air Transport and Air Services Conference, Brussels, 10.6.99.

[51] Claude Rakovxsy (DGIV), 'The EU approach on de monopolisation', 1.5.96 (web site) http://europa.eu.int/en/comm/dg04/speech/six/en/sp96014.htm

[52] Article 90.2, Treaty of Rome.

[53] For details, see: Paul Hitchings, 'Access to International Telecommunications Facilities', in *European Competition Law Review,* Vol.19, No2 (Feb, 98), pp.85-98.

[54] European Commission, Press Release: IP/90/589.

[55] Article 90.2, Treaty of Rome.

[56] See Chapter 5.

[57] See, in particular: Simon Bishop, 'Antitrust Enforcement and the Rule of Law', in *European Competition Law Review,* Vol.19, No1 (Jan, 98). pp.1-3.

[58] For further examination, see Chapter 6, Section 2.

[59] Additionally, it is likely that a firm that does not comply will put itself at a disadvantage, as it may appear to the Commission that it has something to conceal. See D. Woods, op.cit, pp.6-7.

[60] See Article 14, Regulation 17, op.cit.

[61] D. Woods, op.cit, p.7. See Article 15, of Regulation 17, op.cit.

[62] For further information on judicial review, see: L.J. Brinkhorst, L.J. and Henry G. Schermers, *Judicial Remedies in the European Communities: a Case Book* (Deventer,

Kluwer, 1977); Ian McLeod, *Judicial Review* (Chichester: Rose Law, 1993); Michael Fordham, Michael (eds), *Judicial Review Handbook* (London, Wiley Chancery Law, 1994). For a general study on EC law, see: J. Shaw, *European Community Law* (London: Macmillan, 1993), and subsequent editions.

[63] Wilks, op.cit, 1992, p.7.

[64] *Hoffman-La Roche* 1979, op.cit.

[65] *AKZO* 1991, op.cit.

[66] Cases 209-213/84, *Ministère Public v. Lucas Asjes:* [1986] E.C.R. 1425, [1986] *Common Market Law Review,* Vol.3, No.173.

[67] Case 66/86, *Ahmed Saeed Flugreisen and Silver Line Reisebüro GmbH v. Zentrale zur Bekämpfung unlauteren Wettbewerbs E.V.:* [1989] E.C.R. 803 [1990] *Common Market Law Review,* Vol.4, No.102.

[68] See Chapter 3.

[69] D. Woods, op.cit, p.5.

[70] I. Van Bael, 'The Antitrust Settlement Practice of the European Commission', in *Common Market Law Review* [1986], No.23, pp.69-90. N.B. the literature on the use of Comfort Letters and 'Secret Jurisprudence' is extremely limited.

[71] It should be noted that although the Commission may fine undertakings, it has no power to award damages to a complainant, Therefore it is normally in the interests of a complainant to settle a dispute at the earliest opportunity, well before reaching the stage of the College of Commissioners, or Court.

[72] Stephen Wilks, 'Models of European Administration', Conference of the European Group of Public Administration, 2-5.9.92, p.30.

[73] Interview: Professor Steven Wilks, University of Exeter (via telephone), 3.6.98.

[74] I. Van Bael, 'The Antitrust Settlement Practice of the European Commission', in *Common Market Law Review* (1986), Vol.23, pp.69-90.

[75] Wilks (1992), op.cit, pp.30-33.

[76] See: Trevor Soames, 'Opportunities and Challenges in the Aviation's New World Order: The Application of EU Competition Rules to Air Routes from the EU to Third Countries', paper presented at the European Air Law Association, Vienna, Nov. 1998.

[77] For a discussion of comitology, see: Christopher Docksey and Karen Williams, 'The European Commission and the execution of Community policy', in Edwards and Spence (eds), op.cit, 1997, pp.125-153, especially pp.129-130. See also: UK House of Lords Select Committee on the European Communities, 'Delegation of Powers to the Commission, Session 1985-1986, 19th Report, 1986.

3 The Legislative Background to EU Air Transport Policy

Up to this point we have been concerned with theoretical models of interest representation and practical examinations of the changing nature of policy formulation within the EU. This Chapter profiles the legislative background to the formulation process of the Proposal for the ground-handling Directive, and the attempted formulation for a Regulation on slot-allocation. This is in order that the theories and propositions examand in Chapters 1 and 2 can be applied to these case studies.

EU Air transport policy background

The European Commission's Proposal for a Council Directive on ground-handling[1] was formulated in December 1994.[2]

The Council Regulation for the allocation of slots at Community airports was formulated in January 1993.[3] However, the Regulation stipulated that the Council must decide upon a revision of the Regulation by 1 July 1997, based upon a Commission Proposal submitted no later than 1 January 1996.[4]

The issues of ground-handling and slot-allocation are integral to the process of the liberalisation of the EU air transport market[5] since the *Nouvelles Frontières* case,[6] where the European Court of Justice confirmed the applicability of the EC Treaty competition rules to air transport.[7] Prior to *Nouvelles Frontières* the Commission had achieved relatively little in the way of promoting the liberalisation of the air transport market.[8]

The Commission issued a memorandum[9] in July 1979 recognising the need to apply Articles 85 and 86 of the EC Treaty (Competition) to the air transport sector.[10] A Proposal for a Regulation of this nature was formulated, but not adopted by the Council of Ministers. In 1982 a challenge was made by Lord Bethell MEP to the European Court of Justice regarding the Commission's failure to act in the alleged 'collusion' between airlines in the setting of passenger fares.[11] However, this challenge was unsuccessful,[12] and the Commission subsequently stated that it saw no grounds to examine either airline or state activity under the competition provisions of the EC Treaty.[13] A second Commission memorandum[14] was written in 1984 during which time air transport deregulation was taking place in the US.[15] This memorandum suggested that similar market deregulation would be unlikely to happen in the EC because of substantial differences in market structure. However, Balfour has identified the US deregulation as a factor which contributed to a 'change of attitude' within the EC.[16] In addition to US de-regulation, Balfour identifies the conclusion of the Single European Act in 1986 as a contributory factor to air transport liberalisation. The SEA resulted in most related industrial policy having to be formulated by Qualified Majority Vote, rather than unanimity.[17] It also created an impetus towards the completion of the internal market of which air transport was an important element.[18]

Although US de-regulation and the completion of the Single European Act provided *facultative* conditions for air transport liberalisation to take place, it was, however, *Nouvelles Frontières* (a factor also identified by Balfour), which provided the key *necessary* ingredient for the Commission to realise its powers in this area. Indeed, Handley Stevens, the former UK Director for International Aviation has argued:

> ...until the *Nouvelles Frontières* case spelled out the consequences of the Council's failure to act, it was still possible to behave as if the intergovernmental consensus in the Council could oblige the Community to accept the lowest common denominator of change...
>
> ...In the light of the Court's judgement the Commission pressed home its advantage by initiating cases against inter-airline agreements, confident that the agreements would be struck down if the cases ever came to Court. John More, the British Secretary of State [for Transport] piled in too with a threat to apply the Treaty rules through UK law if agreement could not be reached in the Council.
>
> Now that the Council had to act, the Commission could and did insist, in alliance with the more liberal member states,[19] that exemptions from the competition rules of the Treaty would not be granted unless the Council adopted a package which went significantly beyond what the cautious majority regarded as prudent and reasonable. The supranational tendency was clearly in the ascendant over the intergovernmental tendency as the

> Council painfully negotiated the first package at seven meetings between June 1986 and June 1987.[20]

Nouvelles Frontières was paramount to the episode of air transport liberalisation in the EC as it essentially guaranteed the Commission's ability to apply Articles 85 and 86 of the EC Treaty by way of using Article 89 of the Treaty.[21] By using this 'judicial stick' the Commission was able to persuade the Council to reach an agreement on liberalising the manner of fare setting, the easing of rules on 'capacity sharing' and the application of competition rules to air transport.[22] In response the Council adopted three 'packages' of air transport liberalisation, in 1987, 1990 and 1992.[23]

The first package consisted of two Regulations. One implemented Articles 85 and 86 EC to air transport between Member States of the EU (intra-EU air traffic).[24] Additionally, a Regulation was adopted which allowed the Commission to apply block exemptions[25] to Article 85.[26] These block exemptions included the areas of slot-allocation and ground-handling at Community airports, and were set to expire on 31.1.91.[27]

The second package extended the above block exemptions to 31.12.92.[28] The block exemption regarding ground-handling services subsequently expired at the end of 1992 without replacement.[29]

The third package consisted of three 'complementary' sets of Regulations, to extend the scope of the previous two packages.[30] The three areas covered by the Regulations in the third package were:

1. The establishment of common rules for the licensing of air carriers.[31]
2. Free access for Community airlines to all intra-Community routes.[32]
3. Freedom for carriers to set their own fares, without government interference.[33]

The third package also extended the exemption on slot-allocation beyond 1992, later to be replaced with the 1993 Regulation.[34]

Ground-handling

Within the EU the market for ground-handling services has traditionally been closed to competition.[35] Very often it was only the airport or the flag carrier that was entitled to supply ground-handling services. 'Self handling'[36] was rarely permitted, or was reserved to particular airlines on the basis of non transparent criteria.[37] The pricing of ground-handling services was therefore subject to a monopoly situation, where carriers would have no means of improving the quality of the service they received or of negotiating the price they were charged. Subsequently, the Commission stated that the organisation of the ground-handling market did not comply with the principles of the Single Market in meeting the needs of air transport. DGIV (competition), brought attention to complaints made to it concerning the restrictive nature of the ground-handling market by way of a press release in 1993.[38] The Commission also publicly stated that the formulation of a ground-handling Directive was needed in order to 'harmonise' the air transport market across the EU, and to 'ensure the full benefit' of air transport liberalisation, 'which seeks to maintain fair competition between carriers for the activities ancillary to air transport.'[39]

However, when the Commission formulated its Proposal for a Council Directive it also stated that the supply of ground-handling services was subject to a number of 'practical constraints'.[40] Additionally, the 'social repercussions' of market liberalisation needed to be taken into account, the Commission argued.[41] The Commission stated that the Proposal aimed at the liberalisation of the market, but there was also a need for,

> ..a balanced approach which [took] account of the different situations, the requirements of air transport and the interests of airports, carriers, suppliers and workers throughout the industry.[42]

These constraints proved to be 'anti-liberalising' factors in the formulation of the Proposal.[43]

Whilst the Proposal for a Directive was formulated by DGVII (transport), the original complaints regarding this above monopolistic situation were submitted to DGIV by airlines which felt they could benefit from a more liberal regime.[44]

The barriers to entry for the ground-handling market across the EU were, and arguably still are, highly unequal. Several airports (particularly those in the UK), are considered to be 'liberal' in their provision of ground-handling services,[45] whilst other airports are considered to have a 'closed' market. Amongst these airports are Frankfurt, Milan and Athens about which the original complaints to DGIV were made.[46] Rather than acting on these complaints, DGIV passed the views to DGVII to be taken into account when formulating new legislation.[47]

Slot allocation

The International Air Transport Association (IATA) defines a slot as: '...the scheduled time of arrival or departure available, allocated to an aircraft movement on a specific date at an airport'.[48]

Traditionally, slots were allocated by bilateral agreements between the national flag carriers (and thus the governments), of the relevant states.[49] However, 'slot congestion' (particularly at popular airports, and at airports necessary for the successful operation of an airline), results in a situation where carriers which have a relatively large number of slots are at a commercial advantage in comparison to those airlines which do not.[50] Possible remedies for this situation will be examined in Chapter 6. These different possible remedies provide different negotiating positions upon which the interest representation is based.

Path to legislation

The Commission Proposal for the 1993 Slot Allocation Regulation[51] was formulated in 1991. The subsequent report of the Economic and Social Committee[52] took a *laissez faire* approach to slot-allocation, stating that the non-legislative 'guidelines' (which existed at the time), governing the allocation of slots at Community airports were effective, and therefore legislation may not be needed. It emphasised that slot-allocation problems only arise when an airport is 'congested',[53] and this situation only applies to some airports in the Community.[54] Additionally, the report stated that the problems of slot congestion as highlighted in the Proposal were 'settled satisfactorily' by the relevant bodies in the Member States, and therefore the Proposal might have violated the principle of subsidiarity.[55]

The report of the European Parliament Committee on Transport and Tourism argued that the Commission Proposal should include a 'very strict' definition of a congested airport, and a redefinition of what constitutes a slot.[56] This approach was supported in the amendments put forward by the European Parliament. Furthermore, proposed amendments included further research into the reasons for slot congestion.[57] However, these, as well as the vast majority of the proposed amendments of the EP were not included in the eventual 1993 Regulation.[58]

Summary

Traditionally, air transport has been a matter that the Member States (and the 'elite' interests within the Member States), have prefered to be dealt with on a state controlled basis. However, the hand of the Commission in this area was 'forced' by the European Court of Justice in the *Nouvelles Frontières* case. Most EU flag carriers were satisfied with the air transport situation that existed prior to 1986, as the high degree of regulation that existed allowed fares to be kept artificially high.

It is unlikely that the Commission would formulate a Proposal that went against the wishes of these elite interests within the Member States, as the Member States would not allow the Proposal to be passed into legislation at the Council of Ministers. Indeed the Commission did not, and arguably could not, formulate the three packages of liberalisation until the ECJ gave it the legal sanction to force the Member States to do so, a point argued vehemently by Handley Stevens.

The ECJ is, by its nature, more resistant to interest representation as the ECJ does not have to 'answer' to Member States or to their interests. As argued in Chapter 2, a similar (although not quite as intensive) independence is displayed by DGIV, a *quasi-judicial* element of the Commission, charged with upholding the competition elements of the EC Treaty. DGVII (Transport), a legislative element, is more susceptible to interest representation, in particular, representation from 'elites.'

To summarise, the case study of the formulation of a Commission Proposal for a Council Directive on ground-handling is a case study of a Proposal formulated in December 1994. The case study of the attempted Commission Proposal for the revision of the slot-allocation Regulation is a case study of a failed attempt by the Commission to formulate a Proposal, despite being required by EC law to do so. There follows an examination of these two situations in the light of the two hypotheses; hypothesis 1 relating to a reflection of the properties of liberal intergovernmentalism - *a neo-liberal intergovernmentalism* - within the European Commission, and hypothesis 2 relating to the dominance and role played by elites within interest groups.

Notes

[1] Ground-handling is, broadly, any non-flying activity required for the successful operation of an air service. Comprehensive details are contained within Chapter 5.

[2] 'Commission Proposal for a Council Directive on access to the ground-handling at Community airports', COM (94) 590 final, 13.12.94.

[3] 'Council Regulation 95/93 on common rules for the allocation of slots at Community airports', O.J. No. L 14/1, 12.1.93. A slot, as defined in Regulation 95/93, Article 2 (a), p.2, is the scheduled time of arrival or departure allocated to an aircraft movement on a specific date. For comprehensive details, see: Chapter 3 and Chapter 5. For a comprehensive discussion on the legal nature of slots (central to this book), see Chapter 6, Section 2.

[4] Article 14, O.J. L 14/1, 22.1.93, op.cit, p.6.

[5] The terms liberalisation and deregulation are often used interchangeably. Essentially, real deregulation can never exist, as in such a situation no laws would exist to govern an activity. However, an activity such as air transport can be 'more or less' deregulated. The above EC legislation was aimed at creating a 'more' deregulated air transport market. Such a market can also be described as a 'liberal' market. For an economic evaluation of air transport liberalisation in the EU see: P.L. Marín, 'Competition In European Aviation: Pricing Policy And Market Structure', in *The Journal of Industrial Economics,* Vol.XLIII, No.2 (Jun, 1995), pp.141-159.

[6] Joined Cases 209-213/84, Ministère *Public v. Lucas Asjes:* [1986] E.C.R. 1425 [1986] 3 *Common Market Law Review* 173.

[7] See: John Balfour, 'Air Transport - A Community Success Story?', in *Common Market Law Review*, Vol. 31 (1994), pp.1027. See also: Neil Kinnock (Transport Commissioner), 'The Liberalisation of the European Aviation Industry', in *European Business Journal,* 1996, pp.8-13; The Chartered Institute of Transport in the UK, 'Kinnock bids for mastery of Europe's skies', in *Global Transport* (Autumn, 1996), pp.80-81, and Autumn 1995, 'Battle for the Open Skies', pp.48-49.

[8] It is widely argued that the world aviation market is one where a high amount of regulation restricts competition. See: Introduction and Chapter 4. Also see: Christer Jönsson, *International Aviation and the Politics of Regime Change* (London: Frances Pinter, 1987), and K.G.J. Pillai, *The Air Net - The Case Against the World Aviation Cartel* (New York: Grossman, 1969).

[9] European Commission: 'Air Transport: A Community Approach', 4.7.79.

[10] Bernardine Adkins, *Air Transport and EC Competition Law* (London: Sweet and Maxwell, 1994). P.J. Slot and P.D. Dagtoglou (eds), *Towards a Community Air Transport Policy - The Legal Dimension* (Boston: Kluwer, 1989); Trevor Soames (Partner, Norton Rose), 'Access to Airport Facilities and the Impact of Competition Law', paper presented at the Royal Aeronautical Society, 'Slots and Access to Airport Facilities', conference, 21.1.97, p.4.

[11] Case 246/81, *Lord Bethell V. Commission* [1982] E.C.R. 2277 [1982] No.3 Common Market Law Review, No.300.

[12] Adkins notes that the ECJ held that the Commission was under no duty to address a decision to Lord Bethell (Adkins, op.cit, p.5).

[13] Ibid, p.6.

[14] European Commission, Memorandum No.2 on Civil Aviation, 'Progress towards the development of a Community air transport policy', COM (84) 72 final.

[15] A discussion of the US market and its applicability to the EU market can be found in, 'US Airline Deregulation: The Lessons for the European Community', UK Dept. of Transport, Working Paper No.109, September 1989. See also: BH Baltagi, JM Griffin and DP Rich, 'Airline Deregulation: The Cost Pieces of the Puzzle', in *International Economic Review* Vol.36, No.1 (Feb, 1995), pp.245-260; Greg Williams, *The Airline Industry and the Impact of Deregulation* (Aldershot: Abebury, 1994); Alan P. Dobson, *Flying in the Face of Competition* (Aldershot: Abebury, 1995).

[16] Balfour, op.cit, p.1026. See also: KM Gwilliam, 'EEC Transport Policy Issues', paper presented at Rotterdam Transport Centre, Erasmus University, Rotterdam, 1992.

[17] EC Treaty, Article 144.

[18] Balfour, op.cit, p.1027.

[19] Particularly the UK, Netherlands and Denmark.

[20] Handley Stevens (former UK Director for International Aviation), 'Liberalisation Of Air Transport In Europe', London School of Economics, European Institute paper, June 1997, p.49.

[21] See: Chapter 2.

[22] See: Adkins, op.cit, p.8.

[23] Liberalisation 'packages' referenced below. See Balfour, op.cit.
Also refer to: Hussein Kassim, 'Air Transport', in Hussein Kassim and Anand Menon, (eds), *The European Union and National Industrial Policy* (London: Routledge, 1996), pp.107-131; R.G. Brown (EU Economics Studies Group), and F. Sorensen (EU Commission, DGVII), 'New Developments in European Air Transport', paper presented at the Royal Aeronautical Society, London, 1990; Kenneth Button and Dennis Swann, 'European Community Airlines - Deregulation and its Problems', in *Journal of Common Market Studies,* Vol XXVII, No.4 (Jun, 1989), pp.259-282.

[24] (EEC) 3975/87 O.J. 1987, L 274/1. In the *Ahmed Saeed* case, 1989, the ECJ held that competition rules apply to air transport between Member States and non-Member states, in addition to the international air transport between Member States legislated for in the first package. Case 66/86, *Ahmed Saeed Flugreisen and Silver Line Reiseburo GmbH* v. *Zentrale zur Bekämpfung unlauteren Wettbewerbs* E.V.: [1989] E.C.R. 803 [1990] 4 *Common Market Law Review* 102.

[25] Where a certain aspect of a policy is exempt from the EC rules of competition. Block exemptions are needed, for example, to allow airlines to co-ordinate their schedules.

[26] (EEC) 3976/87 O.J. 1987, L 27(EEC) 3976/87 O.J. 1987, L 274/1.

[27] Also block exempted: Capacity planning, revenue sharing, aspects of tariff consultation and Computer Reservation Systems. Commission Regulations (EEC) 2671/88, 2672/88, 2673/88 O.J. 1988, L 239/9.

[28] Commission Regulations (EEC) 82/91, 83/91, 84/91 O.J. 1991, L 10/7. Additionally, the second package applied a block exemption on various rates of cargo air transport. (Council Regulation (EEC) 2344/90 O.J. 1990, L 217/15.) See Balfour, op.cit, p.1032.

[29] Until the 1996 ground-handling Directive.

[30] Michael Niejahr and Giuseppe Abbamonte, 'Liberalization policy and state aid in the air transport sector', in *Competition Policy Newsletter*, Vol.2, No.2 (Summer 1996).

[31] Council Regulation (EEC) 2407/92 O.J. L 240, 22.8.92.

[32] Council Regulation (EEC) 2408/92 O.J. L 240, 24.8.92. This refers to the ability of a Community carrier to operate between any two points in the EU. From April 1997 the Regulation prevented a state from refusing to allow such an activity, although most states allowed such rights prior to this date. However, as certain commentators have pointed out, this 'technical' free market access may be worth little if airlines do not have sufficient access (i.e. access to the slots), at the airports that they wish to operate from. See: Sean D. Barrett, 'Aviation and the Pursuit of Contestability', in *Economic Affairs*, Vol.14 (Feb, 1994), pp.11-18, and, Richard Branson, 'Market Access in the Airline Industry', in *Economic Affairs*, Vol.14, No.2 (Feb, 1994), pp.6-10.

[33] Council Regulation (EEC) 2409/92 O.J. L 240, 24.8.92.

[34] Council Regulation (EEC) 95/93 O.J. L 14/1, 22.1.93; Commission Regulation (EEC) 1617/93 O.J. 1993, L 155/18. Additionally, the block exemption on Computer Reservation Systems was extended, later to be replaced by Commission Regulation (EEC) 3652/92 O.J. 1993, L 333/37. See Chapter 1 for further details.

[35] For further information on ground-handling see Chapter 5.

[36] Where an airline administers its own ground-handling activities. See ibid.

[37] Where the criteria to compete in a particular market are not clear in their form or execution.

[38] European Commission (DGIV) (Press Release), IP/93/714, 5.8.93. See also Chapter 6, Section 2;COM (94) 590 final, op.cit, pp.2-5.

[39] 'Commission Proposal for a Council Directive on access to the ground-handling at Community airports', COM (94) 590 final, 13.12.94, Explanatory Memorandum, p.2.

[40] 'Practical constraints' are considered to be matters such as the availability of physical space for competition, the practicality of duplicating the machinery required for ground-handling and matters of security, safety and the environment.

[41] COM (94) 590 final, op.cit, p.6.

[42] Ibid, p.13.

[43] See Chapter 6, Section 2.

[44] Interview: Rene Smith (EC Relations Branch), UK Dept. of Transport (via telephone), 16.10.96.

[45] Various UK interests highlighted a perceived problem of EC legislation in that a 'liberalising' measure such as the ground-handling Directive, is often only a liberalising measure for those Member States which have a 'restricted' air transport regime. (Interview: Richard Gooding (Chief Executive), London City Airport, London, 9.4.97.) For 'liberal' states such as the UK, such a measure may result in competition that already exists being constrained. This particular problem may highlight the disadvantage of the universal applicability of such EC legislation.

[46] See: European Commission (Press Release), IP/93/714, 5.8.93. See also Chapter 5 for further details.

[47] The reasons for which are examined in Chapter 6, Section 4.

[48] IATA, *Scheduling Procedures Guide* (Geneva: IATA, 1994), point 3, p.5.

[49] See: Peter Reed, 'Flying into trouble', in *New Economy*, Vol. 2, No. 1 (Spring, 1995), pp.9-13.

Hussein Kassim, 'Air Transport Champions: Still Carrying the Flag', in Jack Hayward (eds), *Industrial Enterprise and European Integration - From National to International Champions in Western Europe* (Oxford: Oxford University Press, 1995), pp.189-214.

[50] See: MA Petetaf and R Reed, 'Pricing and Performance in Monopoly Airline Markets', in *Journal of Law and Economics,* Vol. XXXVII (April, 1994), pp.193-213; K. Hendricks, M. Piccione and G. Tan, 'The Economics of Hubs: The Case of Monopoly', in *Review of Economic Studies,* Vol.62 (1995), pp.83-99. See also Chapter 5. Karen Walker, 'Bespoke fortunes', in *Airline Business,* Vol.13, No.1 (Jan, 1997), pp.32-35; JR Burton, CD Zick and RN Mayer, 'Consumer Views of the Need for Government Intervention in the Airline Market', in *Journal of Consumer Affairs,* Vol.27, No1 (Summer, 1993), pp.1-22.

[51] European Commission, 'Proposal for a Council Regulation (EEC) on common rules for the allocation of slots at Community airports', COM (90) 576 final, 30.1.91.

[52] Economic and Social Committee, 'Opinion on the proposal for a Council Regulation on common rules for the allocation of slots at Community airports', O.J. C 339/41, 31.12.91, pp.41-44.

[53] The statistics to prove whether an airport is technically 'full' can be manipulated by all parties and interests. For example, Heathrow, London, is the most congested of EU airports. Several airlines have stated that they have waited for several years to be granted slots from Heathrow (evidence to House of Commons Transport Committee, 10.7.96). Simultaneously, George Younger, former Secretary. of State for Transport stated that there were 25,000 (per annum), slots free at Heathrow (evidence to House of Commons Transport Committee, 17.7.96). This discrepancy results from the fact that an airline wishing, for example, to seek entry to Heathrow, may be offered a number of 'rotations' (where an aircraft completes a take-off and landing), that is too low to operate at a profit. Additionally, the timings of the slot may not be adequate for a carrier to operate at a profit. Thus, the problem of slot-allocation becomes one not of a purely numerical factor, but one where the number of commercially viable slots is of prime importance.

[54] The report highlights the following congested airports: London (Heathrow and Gatwick), Munich, Dusseldorf, Frankfurt, Milan (Liante).

O.J. C 339/41. op.cit.

[55] However, the principle of subsidiarity is itself highly problematic. Whilst there is near universal support for the principle (confirmed officially by the Treaty on European Union), there remains a great deal of debate in air transport (as in many other issues), as to the level of authority (be it regional, national or Community), which should be applied to related issues. For further discussions on the concept of subsidiarity, see: William Nicoll and Trevor C. Salmon, *Understanding the New European Community* (Hemel Hempstead: Harvester Wheatsheaf, 1994), esp. pp.288-293; Michael Newman, *Democracy, Sovereignty and the European Union* (London: Hurst and Co, 1996), esp. pp.123-127; Brigid Laffan, *Integration and Co-operation in Europe* (London: Routledge, 1992), pp.8-9.

For related matters see: William Wallace, 'Government Without Statehood: The Unstable Equilibrium', in H.Wallace and W.Wallace, *Policy-Making in the European Union* (Oxford: Oxford University Press, 1996), pp.439-460, esp.pp.452-456.

[56] The report states that a 'congested airport' should primarily be defined as an airport where an air carrier cannot receive a slot within one hour of its requested time. Furthermore, the report stated that a 'slot' should be redefined as the 'agreed' time of arrival or departure, as

opposed to the 'scheduled' time of arrival or departure. This definition would also have included charter air services, see Amended Article 2 (a), p.7, and Amended Article 4, p10, in, European Parliament, Committee on Transport and Tourism, 'Report on the Proposal for a Council regulation on common rules for the allocation of slots at Community airport', A3-0215/91, 31.7.91.

[57] See: European Parliament, 'Proposal for a Council Directive on access to the ground-handling market at Community airports', with amendments, A3-0215/91, O.J. C. 267/70, 14.10.91, pp.70-82, and (as above with compromise amendments); O.J. No. C 13/446., 20.1.92, pp.446-456.

[58] Regulation 95/93, op.cit.

4 EU Air Transport Interest Groups

Having examined the legislative background to European air transport liberalisation, we shall now examine the particular interest groups to which elite and non-elite interests belonged, and which were involved in the examined case studies of the formulation of the ground-handling Directive, and the attempted revision of the slot-allocation Regulation. This Chapter examines the membership composition of the interest groups, their decision making procedures and their interest representation activities.

For airlines, the vast majority of the members of the Association of European Airlines (AEA), fit Chapter One's description of an elite interest. The members of the European Regional Airlines' Association (ERA) in the main fit the criteria of a non-elite interest. The third interest group to be examined is one which represents airports - Airports Council International (ACI). It is a 'non exclusionary' group in that it represents both elites and non-elites. In addition to the three main interest groups of study, a further category will be examined, which, because of the traditionally restricted nature of the European air transport market, is relatively small in number. This category is essentially the 'non-grouped'; which, in this example, consist of an 'independent' airline, that does not belong to the ERA or AEA, and an 'independent', ground handler.

Association of European Airlines (AEA)

Like Airports Council International, but unlike the European Regional Airlines' Association, the Association of European Airlines is based in Brussels. This factor has great significance in its lobbying strategy and in its ability as a group to influence EC policy formulation. A Brussels location allows interest groups the ability to contact officials and policy makers more easily than interest groups that are not based there. Similarly, it allows the policy makers the ability to contact such interest groups for help, information and advice in drafting Proposals and formulating policy.[1]

> For successful lobbying, interests need to be in Brussels. That's why law firms are here too. The best lobbying is not made by those who jet in and out; it's made by those who are based here.[2]

The AEA stress the importance of the *informal contact* that comes about as a result of being Brussels based.

> It would be strange in a regulated environment to have no regular contact with the regulator. I would say that there is a *continuous* contact, much of which is informal...if you go by totally formal procedures then you are simply building in possible obstructions, in particular... time [obstructions].[3]

AEA members

24 of the 26 members of the AEA fulfil the criteria of an 'elite'.[4] The airlines are, in each of their respective states, the largest in terms of financial turnover and number of employees. They possess the authority, organisation and political importance attached to elite firms and possess the resources to lobby at Levels I, II and III of interest representation. The airlines which comprise the AEA are, in the main, the flag carriers of the European Union Member States, plus the flag carriers of eleven other states.[5] All flag carrier airlines are owned, or were once owned by the state from which they principally operate.[6] The flag carrier has traditionally been a public enterprise and drawn its capital from state funds. It has enjoyed a monopoly market on domestic routes (within its own state) and a 'cosy duopoly' on international routes. The flag carrier has traditionally enjoyed exclusions from national competition law, and privileged access to 'slots' at its state's main airports (Kassim, 1995, p.190).

It was the 'Chicago Convention' of 1944,[7] and the following 'Bermuda Agreement',[8] signed by the US and the UK in 1946, which formalised into treaty a national champion strategy for airlines operating international routes. The former agreement created a situation whereby states enjoyed 'absolute sovereignty' of the airspace above their own territory, whilst the latter provided the model upon which 'bilateral' agreements were negotiated between states (Kassim, 1995, pp.190-193; Cooper, 1947, pp.177-243, pp.331-359; Adkins, 1994, p.13). Built into these bilateral agreements there existed further mechanisms which allowed governments to control their flag carriers. Airlines within the bilateral agreements needed to be registered in one of the states involved in the agreement, thus preventing a 'flag of convenience' type situation.[9] Other criteria required by registering states were so severe that, to all intents and purposes, potential competitors of the flag carrier were excluded from competing. These criteria included specific economic and technical fitness criteria, which would be likely to be appropriate only to the individual flag carrier. Additionally, the criteria stated that the airline must be owned and substantially controlled by home nationals, enabling governments to control what airlines were established within their own state. These criteria were established and executed by the individual state to protect their carriers and prevent competitors from being established and bilateral agreements became the policy instruments used by governments to exercise control over scheduled services (Kassim, 1995, p.193).

The flag carrier airline, and therefore the membership of the AEA, is a product of national champion history. Because these airlines were the 'first on the scene', they are in a near unique position to provide the authority and information required by policy makers. Primarily, the view of the flag carrier carries with it significant political implications.[10] Because of this intimate connection between the flag carrier and the state (or the Member State of the European Union), the view of the AEA is likely, according to DGVII (Transport DG of the European Commission) officials, to carry the support of the Member States, which is ultimately required (by QMV), for air transport legislation to proceed through the Council. As one official within DGVII has argued: '...the view of the Member State usually reflects the view of the flag carrier.'[11]

DGVII relies on the group representation of the flag carriers for the formulation of EC air transport legislation. A DGVII official stated that 'policy issues' can broadly be divided into those that are contentious and those that are non contentious. In a contentious issue (where for example, the views of the Association of European Airlines, and the views of the European Regional Airlines' Association are divided), the larger an interest is, the greater its ability to influence DGVII in policy formulation.

> The airlines of the AEA are simply much larger [than non AEA airlines]. Of course, the bigger you are the easier it is to make your views known to us. Also, larger airlines may be more involved in areas that are not [directly] airline oriented, but which may affect them, such as employee rights and industrial policy.[12]

As for the lobbying resources possessed by the AEA members, the official within DGVII confirmed that airlines 'such as British Airways and Air France', lobby via the AEA and also approach DGVII individually to represent their views. The elites therefore possess the ability to lobby at Level III (directly to Brussels); an ability which the non-elite interests and non flag carriers rarely possess. Additionally, the official in DGVII pointed to the 'know how' possessed by members of the AEA in lobbying activity, specifically the ability to effectively target their representation or 'pull the right levers' of interest representation.

> You may lobby 600 MEPs but this is no good if the Council may overrule them. A small, regional carrier would not have the ability to know where to target its input; [but] an airline such as British Airways would be able to identify the right person to target.[13]

Goals and decision making structure of the AEA

The official objective of the AEA is, 'to serve the common interests of the members of the Association'.[14] Additionally, the AEA stipulates that its objectives include:

> ...representing the interests of its members to the institutions of the European Union, to the European Civil Aviation Conference [ECAC][15] to any other international organisations or association involved in or likely to be involved in issues of interest to AEA members and, as appropriate, to individual governments.
>
> It shall advance co-operation amongst its members on any matter likely to be of interest to the membership as a whole and permitted by the laws of the countries in which they operate, whilst respecting the independence of action of its members individually.[16]

The decision making structure of the AEA is technically based on a simple majority vote by the 26 member airlines. Airlines may officially dissociate themselves from a particular position, and the AEA position must note this disassociation. However, although the AEA gives official permission to its members to act in this way, in reality it will normally attempt to achieve a unanimity of opinion between its member airlines when formulating its position on a particular issue. An official, united position is seen as vital in achieving its goals. The admitted result of this emphasis is compromise, often consisting of 'log rolling'[17] between the airlines to achieve such concessions.[18] British Airways (a member airline), has admitted that the net result of this *de facto* decision making process is as follows:

> There tends to be an AEA position that we can all agree on, but we will put the BA position [independently] without trying to undermine the AEA position.[19]

Therefore, although the AEA possess the ability to produce a simple majority (and arguably a firmer) position, the AEA, as an organisation attempts to avoid this situation at all costs, thereby resulting in a situation where most decisions are produced by a lowest common denominator.

As the case study of the ground-handling Proposal demonstrates, the fact that the AEA can establish a common position by unanimity does not prevent members from acting individually. Indeed, the formulation of this legislation[20] was described by British Airways as, 'the first occasion we had stand-up rows in the AEA'.[21] A common position on this issue was eventually reached, but members individually approached the European Commission, thus undermining the collective position of the AEA. It is arguable that the benefits derived from establishing a common position (including the strength of a 'single voice') are undermined if the group's members individually dissent from an established position.

Structure

The AEA has 23 full time staff at its Brussels headquarters. These posts include the Secretary General, Deputy-Secretary General, a press secretary, and personnel dealing with trade affairs, technical affairs, infrastructure and environment, legal and social affairs, research and information, economics and forecasting, statistics, and finance and administration. Present employees of member airlines are not included amongst the staff of the AEA.

The ultimate decision making panel of the AEA is the 'Assembly of Presidents' of the member airlines (AP). The Presidents' committee (PC), conducts the delegated functions of the AP and provides its working representation. The PC works with two committees to aid it in its function; the 'Legal Matters' committee, and the 'Research and Information' committee, which deals with matters such as aircraft technical specifications. These committees, dealing with such a broad range of issues, feed information into the Public Policy Committee (PPC). The primary purpose of the PPC is to co-ordinate the policies of the member airlines. The bulk of the work in policy formulation is achieved in this committee under the delegated authority of the presidents' committee and the Assembly of presidents. Additionally, the PPC deals with the issue of external relations (interaction with other air transport interests and potential lobbying targets), competition rules and taxation. The PPC feeds the AEA policy to the President's Committee, to represent the AEA to policy formulators.

Table 4.1 Policy making structure of the AEA

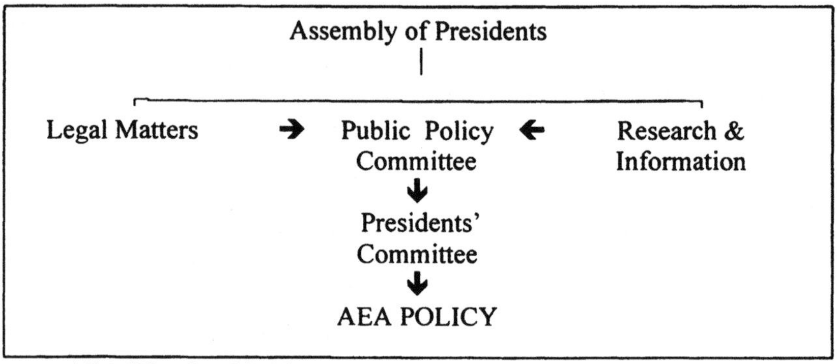

Source: AEA 'Yearbook', 1996

Membership of AEA

New member airlines of the AEA must be approved by the Assembly of Presidents. This factor alone acts as a deterrent against the membership of an airline that Europe's flag carriers may consider a 'threat' to their market positions. Additionally, the membership criteria dictate that members must have been commercially engaged for a period of at least three years prior to application (thus eliminating, or at least postponing 'new entrant' airline membership of the group). Furthermore, AEA member airlines must be of a certain minimum size to be applicable for membership. The AEA criteria state:

> ...one guiding factor [for membership] is the number of aircraft seats operated, which shall be of the order of 3000 seats minimum or for all-cargo operators a minimum payload of approximately 500 metric tonnes.

If such criteria are not met by the major airline in a country, exemption from the above is possible.[22]

Therefore, a measurement of eliteness is specified into this criterion of the AEA, subsequently preventing airlines who are not of a sufficient size from applying for membership[23] unless the carrier is the *major airline* (thus the flag carrier), of a particular European state. The minimum size criteria are subsequently waivered in the case of the flag carrier airlines. Five member airlines fall below the 'minimum size' criteria, yet are members because of their flag carrier status. These airlines are: Adria (Slovenia), Air Malta, Czech Airlines, Icelandair and Luxair (Luxembourg).[24]

Although exact figures are not available, the financial cost of membership is formulated by a complex set of criteria, consisting of the number of passengers and the financial turnover of a particular airline. For example, Virgin Atlantic made enquires about joining in 1997. Virgin was quoted a price of approximately £100 000 per annum. Several of the more 'liberal' airlines argue that whilst these criteria may seem reasonable, it is a disincentive to independent airlines as the funds must be deducted from its own revenue. Airlines that are partly, or fully owned by their host state can expect to have their membership costs covered by their own particular state.[25]

Differences between AEA members in the airline - state relationship

The link between 'flag carrier' airlines, and their host governments, is, on the whole, a close one. For example, a comment made by Gwyneth Dunwoody MP, member of the House of Commons Transport Committee may indicate the close relationship between the UK flag carrier British Airways and UK policy formulators. Dunwoody stated that the Committee's position on the matter under investigation, must reflect the view of British Airways because of the economic importance of the company, in particular, its relatively large number of employees.[26] In the case under discussion (a proposed alliance between British Airways and American Airlines), the Committee supported British Airways' request for the alliance not to be 'modified' in any form by the UK regulatory authority the Monopolies and Mergers Commission.[27]

Moreover, the relationship between the major national air transport interest and its government is apparent when UK civil servants brief transport ministers. Civil servants appear to be guided by two main principles when giving such advice. The first are the legal implications of particular policy options; the second are the 'political realities', of policy options.[28] It is such political realities (in this case, the overwhelming relative importance of British Airways to the UK economy), that place an overwhelming directional 'push' both onto policy makers and their civil servants. O'Reilly of the Air Transport Users Council[29] adds that the UK government will always try to accommodate British Airways, as British Airways has the power to greatly influence national economic policy.[30] This view indicates that, at the domestic level, the views of British Airways are dominant over other air transport interests.

Having said all this, and having painted a picture of an airline 'in bed' with its domestic government, the whole of the European picture must be seen in context. Indeed, when seen in the context of a comparison to the relationship between other members of the AEA and their host governments, the relationship between BA and the UK government appears tenuous to say the least. As we have established, the EU states of the Mediterranean (Portugal, Spain, Italy, Greece), as well as France take a 'less liberal' stance on air transport issues, than the UK, Netherlands and the Scandinavian states. Typically, the flag carriers of the former group of states are state owned; i.e. the government is also the owner of the airline. Therefore air transport policy in those states is normally intrinsically linked with the owner of the flag carrier, i.e. the government of the Member State. British Airways could rightly argue that many of the AEA members are protected from 'market forces', by restricted practices such as direct 'state aid'.[31] This area is one of the most contentious between the 'more' and 'less' liberally-minded members of the AEA. Despite the fact that writers such as O'Reilly and airlines such as Virgin Atlantic cite the link between British Airways and the UK government as one of an extremely close nature, this relationship is *substantially* less 'intimate' than the relationship between 'lesser' liberally- minded governments and their airlines.[32] Therefore, the 'eliteness' of the flag carrier airline is a relative concept. The eliteness of British Airways may be substantial when compared to smaller UK airlines (and by this I do not mean trans-Atlantic airlines such as Virgin, but small regional carriers). However, the eliteness, or 'attachment' between the UK government and British Airways is substantially less than the link between Air France and the French government. Therefore, when seen in the European context, British Airways is no more preferentially treated by its host state than the other flag carriers of the EU – indeed the situation is quite the reverse. Whilst BA and the UK government may occasionally flirt on the dance floor, Air France and the French government have been married for years, and will continue a sordid affair once Air France is divorced and newly married, by way of privatisation, to the French stock market.

As mentioned, British Airways argue that the level of state aid received is a deciding factor if one is to ascertain the extent of the relationship between the AEA member airline and its host government. However, the very existence of state aid underpins the traditional nature of the European air transport system which allows non profitable routes to be operated by the flag carriers.[33] Air France, Aer Lingus, Alitalia, Olympic and Iberia have all been recipients of state aid, the net result of which, according to British Airways, is that these airlines operate in a different type of market to the other members.[34] British Airways claim that such airlines operate loss-making routes, but the financial consequences of this are 'forgotten' as their host state will reimburse the airline's losses. Furthermore, the fact that loss-making routes are serviced prevents 'new entrants'[35] from operating these routes, as the dominant airline, as a result of the state aid, reduces its fares to unprofitable levels, thereby preventing non-elite airlines from entering the market. State-aided airlines also often employ greater numbers than their needs would necessarily require.[36] British Airways opposes the fact that many states continue to subsidise their flag carriers for reasons of 'national pride', and subsequently argues that the attitude of these airlines in their representation at the AEA will not be market-based, but based on other non-market related issues. Speaking about the differences between airlines represented via the AEA, the Manager of competition and regulatory affairs stated:

> There is an EU, non-EU gap; state-owned non state-owned gap; a profitable non-profitable gap [within the AEA].

However, the national significance still clearly attached by many states to their airlines makes this situation most difficult to alter in the interim. The non-profitable, state-owned airlines within the AEA are in the majority, whilst the profitable / non-state owned are in the minority. The table below displays the relationship between ownership and profitability of EU airlines.

Table 4.2 The relationship between profitability and ownership of EU based 'flag carrier' airlines

	Majority Private (√) Majority State (×)	Profit(√) Loss (×)
Aer Lingus	×	×
Air France	×	×
Alitalia	×	×
Austrian	×	×
British Airways	√	√
BM	√	√
CargoLux	√	n/a
Finnair	×	√
Iberia	×	×
KLM	√	√
Lufthansa	√	√
Luxair	√	√
Olympic	×	×
Sabena[37]	n/a	×
SAS	√	√
TAP Air Portugal	×	×

Profit / Loss: Year ending Dec. 1994. Source: AEA 'Yearbook', 96

With the exception of Finnair[38] the EU members of the AEA are profitable when in private ownership and loss-making when in state ownership. Moreover, the 'elite' interest (the flag carrier), is not necessarily the most profitable. Indeed the proportion of non-elites (ERA members), achieving profitability is considerably higher.[39]

The BA manager continues that in most European states,

> ...there is a state airline [which] continues to enjoy massive protection by their government, either by state aid, or by other, less overt, forms of subsidy, [e.g.] advantages at airports, Air Traffic Control charges...none of which we get here.

> The main reason why nothing has changed underpins the nature of the industry. [Other EU airlines] have not had to go through the revolutionary changes we have had here. [The non profitable European airlines should] do what we did; scale back, pull off non-profitable routes, sell non-core assets, and then [they] can grow again. In the meantime, other carriers would come in and compete with them.
>
> [However] that sort of development wouldn't happen [in Europe]. It doesn't matter if they [the incumbent European carriers] lose money...they will still be there, because of state aid.[40]

The question of flag carrier ownership also has a significant bearing upon EC policy formulation. The BA manager argues that the UK,

> ...is generally arguing for the most liberal standpoint. We have one or two allies; the Dutch and Scandinavian countries, Denmark in particular, generally vote with the UK, but it is far from unknown for it to be a 14 to 1 vote [at the Council of Ministers] against the UK.[41]

The AEA is the most exclusive group for airlines. Membership is hard to acquire, and once acquired the financial cost of membership is great in comparison to the membership of ERA (see below). Despite this, the AEA has increasingly become an organisation of contrasts between the various types of 'traditional' flag carrier European airline. The most stark contrast being between those which are state owned (and non profitable), and those which are privately owned (and profitable). Despite this clear distinction, the AEA is the smallest group examined, resulting primarily from the exclusive nature of its membership. Because of this it may be argued that the interests within the group are more 'equal' to one another in the decision making process, than the interests in the groups ERA and ACI, examined below.

European Regional Airlines' Association (ERA)[42]

Unlike the Association of European Airlines, ERA is not Brussels based, but located in Surrey in the UK. Commentators, in particular a prominent member of ERA, have commented on this fact;

> The cost of membership of ERA is next to nothing.[43] We would be willing to pay far more than we presently do to have the job done

> properly and be represented in Brussels...we have argued with them about this for years.[44]

The development of informal relationships (described as vital in the case of the AEA), is limited by ERA's location in the UK. For example, it is difficult for a representative of ERA to ask a member of the Commission or an MEP for a meeting at short notice. As a result, ERA's contact with the Commission is more 'formal' in the sense that meetings need to be arranged substantially in advance. There is, however, a substantial amount of formal contact between ERA and EU legislators covering a range of air transport issues. ERA sits on various committees, provides recommendations and is on 'first name terms' with many Commission officials. Informal relationships are also developed at air transport conferences, of which there are several each year. However, the informality and regularity of contact that is enjoyed by the AEA is impossible for ERA to achieve from its location in the UK.

ERA has a full time staff of thirteen plus a number of secretarial staff. These full-time staff include a Director General, and six managers of infrastructure and environment, projects, air transport policy, administration, external relations and air safety.

Membership and aims

ERA state that its principal aim is to:

> .....foster the development of regional air transport for the scheduled carriage of passengers, freight and mail by encouraging,
>
> • co-operation between members.....
> • co-operation with other groups... interested in the development of regional air transport
> • representation before local, regional, national and European authorities and government organisations.[45]

A regional airline can broadly be defined as an airline using a turboprop or turbofan (jet) aircraft, seating 19-100 passengers. The typical distances flown are described as 'short haul' (i.e. under 1,500 km). In the US a Regional, or a 'Commuter' carrier is officially categorised as an airline whose aircraft have less than 75 seats for a turboprop or 56 seats for a turbofan aircraft (Gleimer, 1996, p.893).

Because the aircraft that regional carriers operate are far smaller than those of the flag carriers, their financial turnover is normally substantially lower. Furthermore, their resources and number of personnel employed are also substantially less than those of the flag carriers. This in turn has a highly significant bearing on the interest representation of regional airlines.[46]

Table 4.3 demonstrates differences between the flag carrier and the largest 'non-subsidiary'*, regional carrier in the UK (including the Isle of Man, Jersey and Guernsey). It serves to demonstrate a comparison between the revenue generated and the employment importance of a flag carrier in comparison to a regional carrier.

Table 4.3 Comparison between UK flag carrier and largest non-subsidiary* regional carrier

	Flag carrier	**Largest Non- subsidiary Regional carrier**
	British Airways	**Jersey European Airways**
No. of Passengers carried annually (millions)	31.73	1.3
Turnover (£m)	12, 142	0, 116
No. of Employees	53, 060	690

Sources: *Airline Business,* Vol.13, No.5 (May 1997), pp.30-33.
AEA, 'Yearbook', 1996, pp.38-62.
Flight International (31 Jul. - 6 Aug. 1996), pp.31-38.
* A non-subsidiary regional airline is one which is not owned by an AEA flag carrier airline (see below).

The flag carrier carries over twenty five times as many passengers, has a turnover one hundred and five times as great, and employs ninety times as many people as the regional carrier.[47]

The relationship between a flag carrier and its government is considerably stronger than that between regional airlines and their respective governments. The regional airline has fewer 'contacts', resulting from the present links (in the case of state owned airlines, such as Air France), and the historical links (in the case of private, once state owned airlines, such as British Airways).

Furthermore, the Chief Executive of a UK regional airline stated: 'We leave European and legislative matters in the hands of the ERA,'[48] adding:

> ..time is the main resource we [regional airlines] do not have. This is the reason that we are members of an association such as ERA; we look to them to make representations on our behalf, as we are too small in our own right....[49]

Whilst AEA members have the ability to lobby their national governments (Level 1), and have the resources needed to lobby individually to the Commission, European Parliament and other EU institutions (Level III), ERA members are Level II (interest group), reliant. Indeed, ERA's air transport policy manager endorses the view that AEA members possess a substantial advantage resulting from their ability to lobby individually as well as via their group.[50]

Regional airlines do not, for historical reasons, possess the organisational and informational resources enjoyed by the flag carriers when they convey their wishes to legislators. This historical advantage, and the substantially high amount of employment generated by the flag carriers results in the situation where the flag carriers are usually the first port of call for legislators. Subsequently, the political importance[51] of the regional carrier is less than the flag carrier. In matters where the views of a regional carrier and the flag carrier are at one, the national government will (unless there are other overriding factors; for example, the opposition of other influential air transport interests, such as airports), support the view of both camps. But where policy differs, the national government will normally side with the flag carrier.

Not only must regional carriers rely on Level II interest representation because of their limited resources, they have fewer personnel to attend ERA meetings when formulating a position. ERA positions are therefore sometimes formed without the knowledge of some of its composite airlines. The position of ERA during the attempted formulation of the slot-allocation Proposal is a particular example of this. The regional carrier British Regional Airlines was not aware that the position of ERA had changed in regards to the issue of 'secondary slot trading'.[52] The issue had not been mentioned in ERA's monthly bulletin, 'Regional Report', although this bulletin is the regular source of information received by all of ERA's members.[53] By contrast, AEA members have a large pool of personnel to attend AEA meetings to discuss various issues and these personnel are normally expert on the political nature of the particular issue under discussion. Conversely, ERA members must usually rely on their 'operations managers' to attend such meetings within the interest group. As a result, it is more difficult for ERA's members to monitor the activities of ERA than it is for AEA members to monitor the activities of their interest group.[54]

Representation

Whilst it is the case that the AEA have a great deal of informal contact with the Commission prior to Proposal formulation, ERA's contact with the Commission appears to follow a more structured and formal approach. One of the key ways for ERA to obtain information concerning forthcoming legislation is the Joint Committee for Civil Aviation.[55] This committee comprises the 'social partners' such as trade associations (e.g. AEA, ERA), unions, and officials from DGVII (Transport). DGVII submits a report to each meeting which will contain information concerning plans for Commission Proposals in the areas of transport. This meeting takes place in two forms; bureau, held each quarter, and plenary, held twice a year. Plenary meetings are more 'general' in their scope. Broad objectives are outlined by DGVII and the various associations. Plenary meetings are notable for the substantial size of attendance; ERA alone is permitted to have up to four people at the discussion table. An inevitable result of this is a 'generalisation of debate' leading to the declaration of broad objectives. The bureau meetings are more thorough in the depth, as opposed to the scope, of the discussion, and ERA will usually send fewer representatives. The meetings last half a day, and ERA's views regarding the various issues are put forward.[56] A key difference between ERA and the AEA is that this is, ordinarily, the first occasion that ERA learns of the Commission's plans for a particular Proposal. Conversely, the AEA is likely to have heard about the planned Proposal at an earlier, informal stage.

Direct contact with the Commission

Contemporary interest representation theory argues that it is the low to medium level officials within the Commission (the drafters of the Proposals), that should very often be the primary target for successful interest representation.[57] The air transport policy manager suggests that he has traditionally had difficulty lobbying these officials within the Commission, and states that such a reluctance by these officials to be lobbied by ERA may be due to a,

> ...civil service mentality; they tend not to want to react to just any voice,[58] we have had difficulty influencing the more 'Eurocratic' level people at a lower level within the Commission.[59]

ERA principally targets the Director of the Air Transport Unit (in DGVII), the Chef de Cabinet of the Transport Commissioner, and the Transport Commissioner himself. Although such individuals might seem entirely appropriate as targets of interest representation, the reality of EC policy formulation dictates that it is the low to middle ranking officials, in tandem with higher ranking officials, who must be targeted for successful interest representation.[60] ERA may, therefore, possesses a 'lower quality' relationship with policy formulators than that of their elite counterpart, the AEA.

Membership

ERA consists of sixty eight 'regular members', forty-three of whom are based in the European Union. ERA provides no categorical requirements for the conditions of membership, although ERA describes regular members as airlines that 'provide two major roles';

> 1. 'The operation of .services from regional points to major international gateway points...'
> 2. 'The operation of services linking regional points.'[61]

Unlike the AEA (whose membership is restricted to airlines), ERA consists of 7 categories of 'associate membership'. The categories cover: manufacturers of aircraft, traders of aircraft, airports, chambers of commerce, non-profit organisations and institutions, aviation consultants and official bodies in regional air transport. This wide membership is illustrative of the 'non-exclusivity' of ERA in comparison to the AEA. There is no minimum criterion, in terms of the number of seats which must be operated by the airline and subsequently no exceptional clauses that excuse flag carriers (the national champions) from having to meet such a criterion. Providing its broad criteria are met, applicants are granted membership. Indeed, the application process consists of a one page form contained within the Association's Yearbook.

The cost of membership is substantially less than the AEA, thus eliminating a considerable barrier to entry. Like the AEA the cost is based on passengers carried. However, the highest rate (for airlines carrying over 1 million passengers per annum), is 8630 Euro (£6000 approximately). Most members pay the rate for 125 000 to 249 999 passengers, that of 3450 Euro per annum (£2400 approximately).[62] In addition to substantially lower member rates than the AEA the method of calculating the membership fee, although similar in that it is based on passenger numbers may be fairer in the case of ERA, as the majority of the member airlines have no ownership connection with their particular state. There is therefore less imbalance between those airlines which can expect their governments to pay the fee and those airlines which cannot.

Decision making

Despite the non-exclusivity of membership *de jure*, the research suggests that a *de facto* exclusivity has developed within the association itself.

ERA's procedure for formulating policy seems to have altered fundamentally in late 1996. This fact, as well as the increased emphasis on the role of the subsidiary airlines of the flag carriers, appears to be largely due to the appointment of an air transport policy manager. Traditionally, all policy formulated by ERA was based upon unanimity. Indeed, a member airline previously stated: 'Mike Ambrose [the Director General], will not form a policy until everyone in that room has said they are happy with it.'[63]

ERA admits that this may have often led to a dilution in the effectiveness of its decision making.[64] The role of the air transport policy manager is to represent ERA to the European Civil Aviation Conference (ECAC), and Joint Aviation Authorities (JAA).[65] The air transport policy manager also represents ERA at the EU institutional level. The appointment led to the Association's structure altering fundamentally.[66]

Until late 1996 ERA consisted of a vast number of 'Working Groups' and 'External Committees.' The former were to disseminate new national and EC policy amongst the members and to provide a forum for discussing such matters. The working groups consisted of four committees (Air Safety, Industry and Commercial Affairs, Maintenance, and Operations). The 'External Committees' consisted of twenty one sub-committees, dealing with matters such as aircraft maintenance, flight crew licensing, noise, security, transportation of elderly and disabled passengers, accident investigation, personnel and working time regulations. These committees were of a highly technical nature. Such a committee structure may have led to an overemphasis on a monitoring of events within the industry rather than forecasting the broader political and legislative ramifications of possible policy formulation. As a result of the vastness of membership and organisational structure, it seems to have become clear to the Air Transport Manager that firm decision making was becoming compromised.

In 1997 the structure of ERA became 'streamlined', and became set out as below.

Table 4.4 New decision making structure of ERA

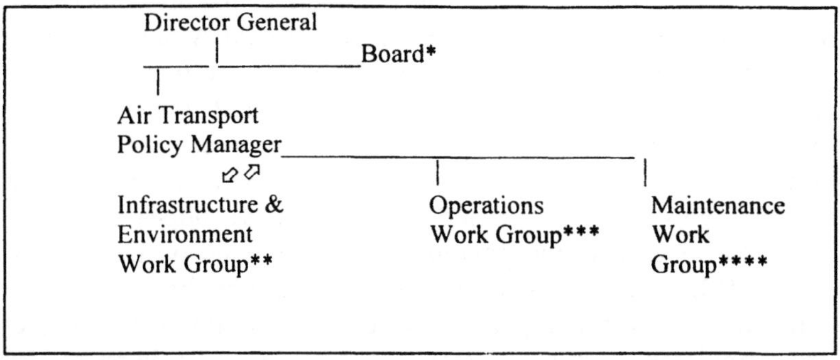

* Elected at annual assembly from airline and associated industry executives. It is charged with overseeing ERA's activities and strategy
**This group has the widest brief, but it can broadly be divided into two. Firstly, it is charged with the formulation of policy positions to present to the EU legislators (hence the direct connection with the air transport policy manager). It is also charged with, 'responding to the major challenges faced in the area of infrastructure, user charges and environmental pressure'.[67]
*** This group monitors the activities of the ECAC, JAA and EU in the technical matters of aircraft operations, and presents policy on operational issues.[68]
**** This group monitors the activities of the ECAC, JAA and EU in the technical matters of maintenance, and presents ERA policy on such issues.[69]
Source: ERA, 'Yearbook', pp.7-40.

Officially therefore, it is the Infrastructure and Environment Work Group (consisting of representatives of all the member airlines), which is now charged with formulating ERA policy. However, research seems to indicate that there are in fact a dozen airlines which *de facto* play the greatest role in the formulation of policy. This method of policy formulation appears to exceed the mandate of the 1997 restructuring set out in Table 4.4. and appears to be based on a minority control by 'subsidiary airlines'.

> We can collect ideas together but we need to formulate a joint policy, and that policy needs to reflect the differences we have; we have a difficulty in achieving that. What I have begun is an informal group of chief executives whom I fax when I need to get a position, and I assemble the answers and produce a position from that.[70]

It appears to be the case that these Chief Executives are primarily from twelve 'dominant' airlines. These dominant airlines are regional carriers who are subsidiaries of various flag carrier members of the AEA. If (as has been stated by officials within the Commission), the AEA have a greater ability to influence policy formulation (as it represents the flag carriers), it may be the case that ERA will carry more weight throughout the Commission (i.e. at all levels), if it can claim to represent the regional carriers which are owned by the AEA members. This greater weight attached to the views of AEA owned ERA members appears to be a contributory factor as to why the position of ERA altered on the issue of slot-allocation.[71] ERA members can be divided into subsidiary[72] and non-subsidiary airlines; the subsidiary airlines seem to dominate when ERA positions are formulated, despite the fact that they are in a minority. ERA would not confirm which airlines have this 'privileged access' to decision making, but an ERA member airline has stated that it is the twelve airlines that are wholly or partly owned by the members of the AEA.[73] Therefore, whilst ERA maintains its non-eliteness *in comparison to* the airlines of the AEA, a significant minority, the twelve subsidiary airlines, have *de facto* control over a group which primarily consists of non-elite interests. However, ERA remains non-elite *relative to* the AEA which, by definition, consists overwhelmingly of elite interests.

This phenomenon of bypassing, 'official channels' in formulating policy, results in decisions being made far more quickly, and probably more effectively, than in the traditional working group structure that existed before the appointment of an air transport policy manager. However, it arguably detracts from the accountability of ERA's decision making process. The table below displays these ERA members (the subsidiary airlines), which are majority owned by the AEA flag carrier airlines.

Table 4.5 Subsidiary members of the ERA

ERA member	AEA member	% owned by AEA member
Air UK	KLM	100 [74]
Avianova	Alitalia	100
BA Regional[75]	British Airways	100
Binter Canarias	Iberia	99.99
Binter Mediterraneo	Iberia	99.99
Brymon	British Airways	100
Cross Air*	Swissair*	67
Delta Air Transport	Sabena	90
Deutsche BA	British Airways	100
KLM CityHopperKLM		100
Olympic Aviation	Olympic	100
TAT European[76]	British Airways	100

Sources: *Airline Business,* Vol.13, No.5 (May 1997), pp.30-31; ERA, Yearbook, 1996/97, pp.56-74 (plus addendum).
* Non EU airline

Regional airlines are non-elites in comparison to the airlines of the AEA. However twelve members are themselves subsidiaries of elite airlines; this has fundamentally altered the decision making structure of the group. Regional airlines are not the largest firms in their representative states in terms of financial turnover or number of employees. Importantly, in terms of interest representation, they do not possess the resources to lobby at Levels 1 (national government) and III (EU institutions); they therefore rely on Level II (group), representation. However, Level II interest representation can be undermined by the dominant interests (the 'subsidiary' regional airlines). It is proposed these carriers, despite the fact that they are a numerical minority, have begun to form an elite *within* ERA.

Airports Council International (ACI)

ACI is a global association consisting of 440 airport operators and 1,200 airports;[77] it is the only international association of airports.[78] ACI is organised into 6 geographical regions: ACI Europe, Africa, Asia, Latin America and the Caribbean, North America and the Pacific. Like the

Association of European Airlines, ACI Europe is based in Brussels. The case made in support of this location to aid its lobbying ability,[79] is considered highly important by ACI - in particular, the ability to easily contact policy makers and officials, and the resulting importance of the informal contact that comes about as a result of its location.

ACI Europe consists of twelve full time staff. These positions are: Director General, head of the European policy unit, head of conference unit, membership development officer, press officer and three policy researchers. Four member of staff have secretarial and translation positions.

ACI Europe has 183 Regular Members; 126 of which are based in the EU.[80] The vastness of its membership may lead to obstacles in the forming of 'firm' positions on matters of air transport policy. It is, therefore, in a similar situation to that which exists within ERA - without a domination by a small group of interests within the overall group, effective decision making would prove highly problematic.

The above studies of the AEA and ERA indicate that EU policy formulators are normally more influenced by the former group than the latter, as the former consists of elite interests. ACI, however, is the sole trans-national group representing airports; it is therefore not possible for policy formulators (such as the European Commission), to exercise an 'interest group preference' in the way that it can if the views of ERA and AEA are opposed to each other.

This examination will attempt to ascertain whether it is the case that all the interests within the group ACI have equal influence over its position formulation. The examination will also attempt to verify whether it is the case that certain interests within the group possess the ability to operate at Levels I, II and III of interest representation, whilst other interests in the group do not possess this ability.

Goals and decision making of ACI

The official objectives of ACI are:

> ....the development of a safe, financially viable air transport system for the benefit of its users and the communities throughout Europe....to promote greater public understanding of airport issues and the important role of airports and the air transport system, and to this end, to work closely with the institutions of the European Union, the European Civil

Aviation Conference (ECAC), and the European Organisation for the Safety of Air Navigation (Eurocontrol).[81]

The role of ACI is therefore twofold: public promotion of the perception of airports across Europe, and promotion of airports to policy formulators and regulators.

Membership

ACI Europe is a non-exclusionary interest group, consisting of both elites and non-elites, that represents 90% of European airports.[82] All airports and airport operators in the European Union that serve international flights are members[83] and all airports, regardless of their size, are eligible to become members. ACI Europe is a branch, officially acting as an independent unit of ACI. Its 183 Regular Members operate a total of 347 airports (as many members, such as BAA, operate more than one airport), and membership is drawn from a substantial area of territory including the former Soviet Union. Membership is of two types; Regular and Associate; the former is for airports (or their operators), and is open to 'airports in geographical Europe with regular or permanent commercial air traffic'.[84] The second type of membership is for those 'seeking airline business'. Associate members benefit from access to Regular Members, themselves potential customers of the Associate members. Associate Members benefit from informational resources as well as being part of a forum in which to sell their products. However, they have no power to vote, and will only be invited to attend 'relevant' Committee meetings[85] at the discretion of the relevant Committee. Subsequently, the primary members of ACI are airports, and the views of airports will, because of the structure of ACI decision making, always prevail over those of Associate members.

Resulting from the non-exclusionary nature of ACI, membership includes both elites (the largest in terms of the number of passengers, usually serving the cities of the greatest importance), and airports which serve a relatively small number of passengers and the less developed regions. For example, if one examines the largest airports and smallest by traffic units,[86] in the UK and Germany, a snap-shot comparison of the variances in membership can be identified.

Table 4.6 A table to show a comparison between the largest and smallest airport / airport operators in the UK and Germany (which are members of ACI)

	Largest Airport Operator	Largest Outside Operator	Smallest Airport Operator	Smallest Outside Operator
UK	British Airports Authority (BAA)	Manchester	Regional Airports Ltd.	Exeter
Traffic units	101 217 908	15 317 438	35 699	227 236
Germany	(no airport operator)	Frankfurt	(no airport operator)	Barth
Traffic units	n/a	48 492 292	n/a	22 730

Source: ACI Europe

Assessing eliteness within a non-exclusionary group

An identification of the 'elites' within ACI is an altogether different task to identifying elite and non-elite airlines. The majority of the 183 airport members are state owned (by the central government of that state), or are owned by a local public authority. The notable exception to this rule is the UK where BAA (the largest airport operator, which possesses seven airports, including the largest, Heathrow), is fully privatised.[87] Subsequently, identifying the ownership of an airport as an indication of eliteness is insufficient as a definitive criterion. However, the evidence below suggests that certain airport interests dominate the policy formulation process both within the group (ACI), and when directing their views to the target of their interest representation (e.g. the Commission). These elites tend to be the largest airports, or their operators. They are the airports which are of the greatest economic importance to their governments and are usually the oldest commercial airports in their particular states. There is, ordinarily, a 'reliance' by governments on these elites, similar to the reliance which exists between governments and their flag carrier airlines, when formulating policy.

Preliminary evidence of this eliteness within ACI is contained in the ACI official briefing paper prepared for DGIV and DGVII of the Commission (Competition and Transport Directorates, respectively), on the draft ground-handling Proposal. Out of 126 EU member airports only 15 were consulted; these were highlighted in the document. Each was the largest airport within each Member State, with the exception of Germany and Italy, where the two largest airports in each state were consulted.[88] Indeed, the document made it implicitly clear that it was representing the largest airports, whilst carefully avoiding any outright statement of this fact. It subsequently attempted to provide a 'group legitimacy' (i.e. that it was reflective of the views of airports across Europe), yet the document was clearly weighted in favour of the largest players.

Decision making and structure

Regular members receive documentation (such as the monthly and annual traffic reports of all ACI members), a monthly news bulletin, press notices and the official positions of ACI on policy matters. Regular members are also entitled to use the ACI logo on company stationary and publications.

In terms of group decision making, regular members are eligible to place their directors and managers on the ACI Europe Committees. The Board of ACI Europe, which meets quarterly, is *de jure* the body which formulates positions.[89] The nine committees which form these positions report to this Board. Eight of the committees are 'technical' and deal with non-contentious issues that the vast majority of members would ordinarily agree upon.[90] However, by far the most important committee, in terms of decision making, is the ACI 'European policy committee'.[91] Members of the European policy committee are not specialists on the particular issues of concern (although they will have some knowledge of the issues, for example, slot-allocation or ground-handling), but they are 'tuned into the political issues', and 'see the political aspects of what might happen',[92] in the words of the head of the European policy unit. The vast majority of negotiation happens at this level, and the Board of ACI appears to be a 'rubber stamp' for decisions made in this Committee. Additionally, evidence suggests that there is a dominance of a number of airports (or operators) within this Committee.

> The European policy committee of ACI is the main committee for making positions - the others are just technical. There are usually about 30 representatives on the Committee, as many airports do not have the time

> to send representation. 20 of them you can describe as 'inactive'; 10 get very involved and direct the decision making process. These are the larger airports such as ADV,[93] BAA, and Airports de Paris.[94]

Non-elite airports lack the time resources to effectively represent themselves within the ACI decision-making process. Subsequently, not only are non-elites handicapped at Levels 1 and III, but they are additionally handicapped within Level II. All 183 regular members are eligible to attend the European policy committee; however only approximately 30 attend the meetings on a regular basis, and only 10 have a substantial input.

A member of ACI that can be classified as an elite stated:

> ...as with many trade associations, there are those who are there to work, the movers and shakers, and...there are members who are there for the ride, because it appears good to be in such an organisation.[95]

A non-elite member stated that the lack of time resources was a handicap in attending meetings to discuss EC policy formulation issues;

> ...the big companies such as BAA have permanent representatives in Brussels,...spend a huge amount on lobbying and have very effective lobbying machines.[96]

The Chief Executive of UK airport London Luton is the only member of personnel at the airport suitable to represent his airport at such meetings, whilst organisations such as BAA have staff to deal specifically with such matters. Luton, he stated, does not have the resources to undertake such an approach.

However, as well as lacking the resource of time to attend such meetings, there appears to an acceptance amongst airports and associated interests that the elites will inevitably dominate the policy formulation process.

> ...whilst our views are listened to, we do not pretend that we have the influence of the large airports. Part of this is due to the fact that BAA can say: 'Let's be sensible; we're a large part of UK GDP'. When it comes to this sort of lobbying, at Westminster or in Brussels, the bigger you are the more clout you have...the big airports have such enormous leverage on the economies of countries, let alone regions, so of course they have to be taken notice of.[97]

An air transport solicitor added, that it is always the case that the larger the interest, the greater its ability to lobby, whether to the government, or to the European Commission and other policy makers.[98]

Subsequently, the elite airports in the case study of the ground-handling Proposal and the attempted slot-allocation Regulation Proposal are the largest airports (or airport operators) in each of the Member States – these airports are, in the main, the same airports identified in the ACI report of 1994 (see above). Furthermore, in the case study of the formulation of the ground-handling Proposal, the evidence suggests that certain elites *within* ACI dominated its position formulation process.

ADV (Arbeitsgemeinschaft Deutscher Verkehrsflughafen)

The evidence suggests that the dominant interest in the formulation of the ACI position on ground-handling was the national interest group that primarily represents the German and Austrian airports - ADV.[99] ACI arguably fell to a lowest common denominator position when forming a position on the ground-handling Proposal largely as a result of the pressure from ADV.

ADV is not the only national association of airports that is represented at ACI. For example, two associations represent French airports, ALFA-ACI *Aeroports de Langue Française Associés á ACI* (the Association of French Language Airports of ACI), and UCCEGA *Les Aeroports Français* (the French Airports). Italian airports are represented by *Assaeroporti* (the Italian Airports Association), and the UK is represented by the AOA (the Airport Operators Association).[100]

However, as a group, ADV is substantially different from the above national groups, and ADV is by far the strongest voice within ACI. This is due to several factors in common with all German airports. The defining characteristic between ADV, and the representation of other EU airports is that ADV, although a group representing all German airports, is extremely precise in its interest representation and internal policy formulation. The other EU Member States have a 'fragmented' approach by their airports (even where a national interest group exists).[101] It is normally the largest airport (or group) in each Member State that dominates that state's representation at ACI. ADV's largest member is Frankfurt Airport. This airport and Dusseldorf Airport, together with ADV are all members of the ACI European policy committee. However, the views of these two airports and ADV are co-ordinated well in advance of ACI's European policy committee meetings.

When questioned about the role of ADV (in the case study of the ground-handling Proposal), the Head of Airport Strategy at BAA appeared to argue that it was the dominance of ADV which led to a situation where the position of ACI became a lowest common denominator;

> When we as a company are faced with a situation where one of our partners at ACI faces serious consequences [as a result of the possible legislation] - what do we do?...we could approach it from the point of view that ADV are just being unreasonable [and] unrealistic; they're going to have to change anyway and we'll fight them all the way and ensure that ACI represents a 'liberal' view to the Commission....however there would be no point doing that....
>
> ...there was really little point falling out with them, knowing how strongly they felt about the issue of ground-handling; we understand the potential impact on their business, as well as the social implications; so what we did...fairly consciously at the time, was to allow airports such as Frankfurt and others with a very strong view to lead the position in ACI..there was, as a result, a type of lowest common denominator approach by ACI.[102]

This view is supported by the Chief Executive of London City Airport;

> ACI struggled as....the German airports were most vociferous in opposing the [draft] Directive...we knew that there would be changes but it would not be a catastrophe, whereas at Frankfurt they would lose 40% of their income.....ACI was a bit neutered [as a result].[103]

The influence ADV possessed resulted and continues to result, from two broad factors;

1. The similarity of the ownership structure of German airports and their integrated nature with other German institutions.
2. The decision making structure of ADV.

1. In Germany each airport is considered an independent entity, yet each is either owned by a combination of the Federal government, the Länder (region), and the City.[104] Alternatively, some German airports are owned by only by the Länder and the City.[105] Additionally, these individual Länder,

the cities and the German Chambers of Commerce are all members of ADV, adding their substantial 'weight' to the ADV decision making process, as well as to the representation of ADV. Subsequently, governmental institutions and the other interests that they may represent (such as those involved in job creation and preservation), all become involved in airport interest representation. As the Länder are airport owners, in addition to being members of the Bundesrat (Federal Council), a situation exists where the airport owners are also parliamentarians. Also noted is the role of the German trades Union, ÖTV (Öffenntlicher Transport Verband); the, 'Public Services Transportation Union'. ÖTV is not a member of ADV, and the Chief Executive of ADV states that during the formulation of the ground-handling Proposal, ADV and ÖTV, 'were not in the same boat'.[106] However,

> ADV and ÖTV had similar interests and we co-ordinated our efforts. Our views were not identical but we co-ordinated and informed each other of our approaches. Generally, the airport management were more in favour of competition than the union...however (like the unions), it was in the interests of airport management to maintain jobs.[107]

Moreover, the sharing of ÖTV's broad objectives and its subsequent support seems to have strengthened the position of ADV in the case study.

ADV makes direct representations to the German government, the Commission (both DGIV and DGVII), European Parliamentarians (such as the *rapporteur* of the Transport and Tourism Committee), as well making representations via ACI. In the case study of the ground-handling Proposal the Bundasrat passed a resolution supporting the stance of ADV.[108] It is understood that this was the only example in the case study where a Parliament (either national or regional), lent such support to an interest prior to the EC legislation.

2. ADV has an extremely precise mechanism for formulating a position. Importantly, it has a procedure within itself for a form of qualified majority voting, rather than using unanimity to formulate its positions; this results in a far stronger application of its position.[109] When an issue becomes relevant, the federal government will ask for a view from ADV. ADV consists of several 'expert' committees (such as the commercial committee and the technical committee), and these committees advise the prime policy making committee the 'legal committee', consisting of the 'Chief legal advisors' from the German airports. They form a common view using a

weighted majority (or a form of QMV),[110] where necessary, after receiving advice from the expert committees.[111] This common view is placed on the agenda of the chief executives to discuss at their meetings, which gather 4-5 times a year; this meeting formalises the position agreed at the Legal committee. Once a position is formulated, ADV is in a position where it can represent all the German airports without any dissension. The chief executive of ADV subsequently has a mandate to speak on behalf of all German airports. The chief executive of ADV stated:

> ...we have a very precise mechanism to come to a view-point...there are not any comparable organisations such as ours, which can formulate the views of all national airports, as we do; it does not happen in other countries...there are, of course, airport associations (such as the Airport Operators Association in the UK), but they play an entirely different role.
>
> In a centralised state, such as the UK, you have BAA, Manchester etc., all speaking for themselves. [Here] an airport such as Frankfurt is big enough to speak for itself...however, when it comes to representation and public hearings its ADV that goes to the Parliament and represents all the airports.[112]

There is a 'culture' of non dissension once an ADV position as been established and airports subsequently undertake to defend the view that ADV expresses. ADV differs from the majority of interest groups in the 'strength' of the relationship that exists amongst its members. The Chief Executive of ADV stresses that *Arbeitsgemeinschaft* means 'working together'.[113] Once an agreement has been made, members are bound to defend that decision even if they do not agree with it.[114]

The Head of Airport Strategy at BAA admitted, 'ADV has a far more formalised structure [than other groups]...it is more like a half-way house between a trade association and a single company'.[115]

Upon examination of Table 4.7 a pattern begins to emerge of the domination by ADV of the European policy committee of ACI; ordinarily, it is the main airport operator in each state that is the dominant player from that state at ACI; subsequently each state's position faces potential fragmentation of the view of the main operator (e.g. BAA) diverges from the view of another operator or individual airport (e.g. Manchester). However, for the German and Austrian airports, all are individual members of ACI, yet all their actions are co-ordinated through ADV.

Table 4.7 A table to demonstrate the ownership structure of EU airports (which are members of ACI)

Germany and Austria, (plus Zurich) :	All airports members of ADV; no ownership by an operator
Denmark:	1 operator of 2 airports (Copenhagen Airports Ltd), plus 2 other airports
Finland:	1 operator of 25 airports
France:	1 operator of 3 airports (Aeroports de Paris), plus 37 others
Greece:	1 airport (Athens), plus others
Ireland:	1 operator of 3 airports, (Aer Rianta)
Italy:	1 operator of 3 airports (SEAP), plus 14 others
Luxembourg:	2 airports
Netherlands:	5 airports
Portugal:	1 operator of 7 airports (Aeroportos e Navegacao Aerea), plus 2 others
Spain:	1 operator of 40 airports (Aeropuertos Espanoles y Navegacion) Aerea (Aena)
Sweden:	2 operators; 1 of 19 airports (Civil Aviation Administration of Sweden), 1 of 32 airports (Svenska Regionala Flygplats Forbundet)
UK:	2 operators; 1 of 7 airports (BAA), 1 of 2 airports, (Regional Airports Ltd.) plus 20 airports

Source: ACI Europe

Because of the nature of ADV, its extremely precise method of decision making, and its 'collective responsibility' once a decision is made, ADV has the most powerful single voice in EU airport policy formulation, in addition to the most powerful voice within ACI, as Chapter 5 will indicate.[116] The Head of Airport Strategy at BAA stated, '..when everyone in a trade association has the same view they are very powerful vehicles for representing the interests of the industry'.[117] ADV combines this collective responsibility, and 'guaranteed' support from the German government to its substantial advantage. In the case of the ground-handling Proposal, the Chief Executive of ADV stated:

> ...Germany is one of the important countries; the Commission did not want to have a confrontation with the German government and run the risk of the whole Directive failing, so they made sure the final Proposal was more 'flexible'.[118]

The Chief Executive added that he had a substantial personal conversation with the Competition Commissioner at that time in order to put the view of ADV.[119] However, he also knew (and therefore so did the Competition Commissioner), that he had the support of the German Bundasrat at the time of the conversation - no other interest had such a powerful ally.

ACI is a non-exclusionary group. Its membership consists both of elites (who possess the ability to lobby at the three levels of interest representation), and non-elites (who are limited to Level II representation). However, *effective* Level II representation is restricted to those that play an 'active' role on the European policy committee; the key decision making body of ACI. The most effective member of the European policy committee was, and arguably still is, the German airports interest group ADV, as a result its highly effective decision making and subsequent interest representation within ACI, at the level of the state, and at the level of the Commission / European Parliament.

Miscellaneous interests

The final group of interests can be described as the 'miscellaneous' interests whose influence must be noted prior to an examination of the two case studies. This category has only two interests to be examined. Neither of the interests is a group but both are firms in their own right. The category is extremely small in number, resulting from the extremely restricted nature of the European air transport until comparatively recently.[120]

Interest 1

The first interest is one of only several 'independent' scheduled airlines.

Until comparatively recently, very few airlines in the European Union, which were not members of the AEA or ERA, operated scheduled services. One airline which was an exception to this rule was the UK based airline Virgin Atlantic which began operations in 1984.[121] Virgin shares the characteristic of the majority of ERA airlines in that it has never been state owned. However, because of the size of its aircraft and the routes which it operates, Virgin is well outside the scope of even the very broad criteria required for ERA membership. Virgin is also not a member of the AEA. Virgin is however a member of the interest group, *Association internationale de charter aerian,* the Association of Independent Air Carriers (ACE). Virgin is, however, unique within this group as the other 14 members of ACE are charter airlines.[122]

Because of this, a highly-placed official within Virgin stated, 'ACE is unready, unprepared and unable to represent the interests of independent airlines in Europe.'[123] The situation of concern to Virgin is not that the interest representation of ACE results in a lowest common denominator of interest representation *per se*, but rather a situation exists where the views that ACE attempts to put forward to policy formulators are not in the immediate or long term concerns of Virgin. Virgin feels that the main benefits to it arising from membership are of a technical nature; for example, forums exist within ACE to discuss safety, engine maintenance, ticketing, and other matters where expertise can be shared.

Virgin is unique as it is the only non-elite interest examined to actively operate at Levels 1 and III, thus largely bypassing Level II (i.e. ACE). The matter of 'interest group dependence' that may adversely affect members of ERA cannot be applied in the case of Virgin, as, despite being what can reasonably be classified as a non-elite interest, Virgin conducts the vast majority of its interest representation activity via Levels 1 (to the UK government), and III, (to the European Commission). It may also have a bearing on the unique role that Virgin plays in policy formulation, particularly in the case of the Proposal for the slot-allocation Regulation, where the views of one arm of the Commission, DGIV, far more closely reflected the views of Virgin than DGVII (see Chapter 6). The Director of government and external affairs at Virgin has stated:

> We are not major participants in the major industry bodies....we are not members of the AEA, although we are large enough to be so. We are partial members of ACE, which really represents leisure airlines, so we do not really have a voice through the major lobbying groups. Our voice [is also] rather small. It is also difficult for us to afford representation.[124]

However, because of the uniqueness of Virgin's position, it is sometimes invited to Commission meetings that no other *individual* airline attends. For example, there have been occasions (such as a Commission discussion on the slot-allocation Proposal), where AEA and ERA representatives were present, together with Virgin, but no single flag-carrier or regional airline was present.[125]

Virgin currently has two full time staff that 'lobby' for the airline; a Director of government and external affairs, and a manager of 'aeropolitical' affairs. Virgin also employs two law firms for this purpose; Harbottle & Lewis, based in the UK and Wilmer, Cutler & Pickering, based in Washington DC. Unlike British Airways, however, Virgin does not have any effective representation in Brussels.[126]

Interest 2

Interest 2 is an independent ground handler; one that is not owned or controlled either by the airport at which it operates, or by an airline.

There are extremely few independent ground-handlers in the EU, primarily because this has also been a restricted market, which had, and continues to have, very high barriers to entry. By far the largest independent ground handler is Servisair, based in Manchester. Servisair supplies a supervisory service to 39 airports outside the UK,[127] yet provides a 'full service'[128] at only four airports outside the UK. The general manager for business development at Servisair states:

> Although we have over 50% of the independent ground-handling market, we are not a big company by any means.[129] It shows you how small the other ground handlers are.[130]

Indeed, the vast majority of the other independent ground handlers in the EU, with the exception of Belgavia, based at Brussels Airport, operate at airports in only one country. Servisair and Belgavia are the only two independent ground handlers which are Associate Members of ACI.[131] Because of the constitution of ACI (see above), neither of these firms is eligible to sit on the European policy committee (unless invited for a specific meeting), and therefore have little direct input into forging an ACI position.

The main voice for interest representation of the independent ground handlers is the Independent Ground Handlers Association (IGHA). However their largest member (Servisair), substantially downplays the role and effectiveness of the Association in comparison to other air transport interests.

> [For example] BAA and ADV are extremely well resourced (sic) with very deep pockets; they can lobby and formulate opinion far more successfully than independent handlers....they have great lobbying power with concentrated, vast resources, and a lot to lose by not succeeding. They have very qualified people to do it [lobby] for them.

> Then you have a bunch of 'enthusiastic amateurs'. [Although] independent handlers are in a group, it is still a group of disparate elements coming together every 3 months to discuss issues whilst the German Airports set up a department to discuss it and already have the ear of the German government by their very nature...so I think the lobbying power will always rest with the mighty......[132]

The inclusions of Virgin and Servisair in the case studies are relevant as both broadly fulfil the criteria of being non-elites.[133] Both are also very sceptical of their respective interest groups. This situation may be due to their uniqueness within their respective groups; Virgin as the sole scheduled carrier in ACE and Servisair as the only independent ground handler of any substantial size.

Non EU membership of above groups

Not addressed in detail in the above information is the issue of non EU membership of the examined interest groups. AEA has 11 non EU members, ERA, 13 and ACI Europe, 57. The fact that a substantial proportion of the membership of the groups derive from non EU member states (AEA: 26%, ERA: 19%, ACI: 31%), may have a bearing on their decision-making procedure. However, according to the Head of Airport Strategy at BAA, the overwhelming bulk of work conducted by ACI deals with EC legislation. This is described as of 'remote' interest to members from non EU states, unless they are affected by the legislation (such as EEA members from Iceland and Norway) or if they are an 'aspiring member' of the EU. This may be a contributory factor as to why not all members are 'active' on committees such as the European policy committee of ACI.[134] If one looks at the 'elites' which regularly sit on the European policy committee, all are from EU states or states which are 'first on the list' for EU membership.[135] Furthermore, the twelve dominant carriers within ERA are all EU members.[136]

Summary

The non-exclusionary group, Airports Council International, although theoretically possessing an equality between its composite members in decision making, seems to give a greater weight to a small selection of its members; these are the members that are categorised as elites. Additionally, from these elites, it is the German airports group, ADV which carries the greatest weight in ACI policy formulation. ACI and ERA are groups where the elites within the group dominate. These two examples support the Greenwood hypothesis that, for a group to be effective, elites must be able to dominate the formulation of group positions (Greenwood, 1994, p.184). The decision-making within the group, the European Regional Airlines Association is, however, unique. By definition, airlines must be smaller than a particular size in order to become members of this group. However, the criteria for membership do not exclude the membership of the subsidiaries of elite carriers, as there exists no rule within ERA to exclude subsidiary airlines. Indeed the presence of subsidiaries within ERA seems to be encouraged by the air transport policy manager precisely because of their access to the elite interests via their ownership structure This number of subsidiary airlines within ERA amounts to less than one fifth of the total membership; however, despite the fact that they are a minority within the group, the evidence presented in Chapter 5 suggests that, in the case of the slot-allocation Regulation, these subsidiary airlines had an enormously disproportional advantage in comparison to the other 56 member airlines. Like ACI, the members of ERA all theoretically have equal access to the decision making procedure. Conversely, the evidence suggests that upon the appointment of an air transport policy manager, a situation was created where 'elites' were identified within a group which is mostly non-elite. Thus, the Greenwood hypothesis has been extended not only to groups representing elites and non-elites but to groups that, are themselves, non-elite.

Because of the relatively small size of its membership, the exacting elite criteria for membership and the subsequent extreme difficulty and expense in acquiring membership,[137] the AEA appears to be the most egalitarian group *within* the confines of its own membership. This results from the fact that the vast majority of the members of the AEA are elites. The views of every airline, even the smallest amongst the AEA, are likely to carry substantial weight within the group.[138] Having said this, a prominent member of the European Parliament's Transport and Tourism Committee seems to downplay its significance.

> ....[The Secretary General of the AEA] seems completely unaware of the difference of views [among] the airlines of the AEA. The AEA is completely ineffective as a lobbying organisation as they all individually

> present different messages. The airports on the other hand come with a single voice and present a single position through ACI.[139]

Perhaps most tellingly, a former highly ranked employee at BA stated:

> BA sees the AEA as an organisation that is useful only for collecting statistical data.[140]

This chapter has identified elite and non-elite airlines and airports (as well as an independent ground handler). It has not addressed the issue of whether an elite airport or an elite airline would possess more influence in a lobbying situation. The case studies below will attempt to verify this, amongst other things. A clue to the answer of this question was provided by the Head of Airport Strategy at BAA, who argued that it may well depend upon the issue at stake. For example, in bilateral negotiations between states on air transport, the flag carrier may well be the dominant interest in relation to its own government. In matters relating to employment (a substantial factor in the formulation of the ground-handling Proposal), airports may possess more influence over their national government. Furthermore, the relative level of influence may also depend on which section of the Commission is promoting the issue.[141] In the case studies below, it is DGVII (Transport), and DGIV (Competition). DGIV, by its nature, will take the view it sees as 'pro-competition.'

In addition to the relative lack of resources of non-elites, and the relative lack of their political importance, a 'barrier' to their participation in interest representation may stem from the fact that certain issues do not affect an airline, or an airport, which is of a relatively small size. For example, the slot-allocation situation is one where several regional carriers have traditionally not been affected as they did not seek access at 'congested' airports.[142] However as regional carriers have expanded their operations and have begun to seek such access, the issue becomes one of growing importance to them.

The UK based airline Gill Airways is an example of this phenomenon. As this issue has become one of concern, the airline has had to consider lobbying strategies in order to improve their own position in policy formulation.

> We've only recently started flying to Düsseldorf [a slot-constrained airport]. We plan to start flying to Paris, [Charles de Gaulle, also slot constrained] thus slot-allocation is only now becoming an issue. As we

> cast our net further and wish to develop services, we now have to think about how we should lobby. Should we form lobbying alliances with other airlines, that share our objectives? What gains can we get from ERA? It's only now that we have to think about that.[143]

Therefore non-elite interests as a result of their size may be limited not only in their lobbying capability, but also by their lobbying necessity. Arguably, the 'lobbying necessity' increases as airlines (or other businesses for that matter), develop their activities.

Finally, research also indicates that there are, broadly, two views which separate the seemingly 'liberal' from the 'non-liberal' in EU airport policy. Just as the airlines from the UK, Netherlands and Scandinavia have been 'liberalised' in that their flag carriers are in private ownership (although maybe still enjoying national champion status), the airports from these states take a different philosophical view on the role of an airport, in comparison to the view of other EU states. France, Germany, and the states of the southern EU argue that airports are pieces of transportation infrastructure and therefore legislation should be formulated based on this particular premise. In the formulation of the ground-handling Proposal, ADV emphasised this fact to the Commission. Indeed, ADV stated that the proximity of German airports (all are closer than 150km to the next), makes them an integral part of the transportation infrastructure needed to facilitate the EC internal market. 'An airport is a transportation facility [and] should therefore be used in the optimum way.....the Commission accepted this....'.[144] Conversely, London City Airport in the UK (a fully privately owned airport), stated:

> What is happening is *airline* deregulation and increased *airport* regulation. What is presented as a policy of liberalisation, is liberalisation only at one level - there is a danger that regulation is shifted from one side to the other - that strikes me as being curious. You might argue that airports are natural monopolies. [However], in London are 5 airports which have competition between them...the days when airports were a monopoly have ended. This is not recognised either by national governments, or by the EU.[145]

Notes

[1] A representative of Manchester Airport confirmed that it may be a substantial advantage for a firm to be located in Brussels. She stated that the location provides a 'presence'

(although, she admits, a costly one), which substantially aids a firm in its representation. Interview: Louise Congdon (General Manager, Strategy), Manchester Airport (interview in London), 8.4.98.

[2] Interview: Scott Crosby (Solicitor), Kemmler, Rapp, Böhlke and Crosby, Brussels, 17.4.97.

[3] Interview: Kees Veenstra (Deputy Secretary General), AEA, Brussels, 16.4.97.

[4] The exceptions to this rule are the airlines British Midland, of the UK, and CargoLux, of Luxembourg. These two airlines are the only ones that the description, 'national champion' does not apply to. The latter airline is the only member that deals exclusively with the transit of cargo. However note well the comments by Jordan on interest group 'legitimisation (Chapter 1, f/n 41). Jordan ascertains that the 'legitimacy' of such groups is increased when smaller members are allowed membership.

[5] EU member airlines: Aer Lingus (Ireland), Air France, Alitalia (Italy), Austrian Airlines, British Airways, British Midland, Finnair, Iberia (Spain), KLM (Netherlands), Lufthansa (Germany), CargoLux, Luxair (both of Luxembourg), Olympic Airways (Greece), Sabena (Belgium), Air Portugal. Non EU member airlines: Adria Airways (Slovenia), Air Malta, Balkan Bulgarian Airlines, Czech Airlines, Cyprus Airways, Icelandair (Iceland), JAT Yugoslav Airlines, Malev Hungarian Airlines, Swiss Air, Turkish Airlines. Scandinavia Airlines System (SAS), was created by an amalgamation of the airlines of Sweden and Denmark and Norway, the latter of which is non EU.

[6] Although there are some notable exceptions to this rule. SAS, was the only 'multinational' airline to exist within Europe. Also, Olympic Airways differs from the norm in this respect. Olympic was started as a private concern by Aristotle Onassis (albeit on receiving the state-owned shambles of an airline TAE). Olympic was eventually taken over by the Greek government in 1975. See: Nicholas Fraser, Philip Jacobson, Mark Ottaway and Lews Chester, *Aristotle Onassis* (London: Book Club Associates, 1978).

[7] The original Chicago Convention can be found in the United Kingdom Command Papers, No.6614, 1945. A verbatim text can also be found in: Appendix A (Part 2), in Peter Martin and Elizabeth Martin (Shawcross and Beaumont Solicitors) (eds), *Air Law*, Fourth Edition, Vol.2 (London: Butterworths, 1998), pp.A91-A118.

[8] 'Agreement Between the Government of the United Kingdom and the Government of the United States of America Relating to Air Services Between Their Respective Territories'. The original agreement can be found in United Kingdom Command Papers, No.6747, 1946. Verbatim text, published in 1947, in *The International Law Quarterly*, Vol.1, No.1 (Spring, 1947), pp.286-289.

[9] This situation is where, for example, an airline registered in Spain operated principally from the UK. This situation is widespread in the shipping industry.

[10] Note O'Reilly's comment below exemplifying the 'control' that a flag carrier airline may have over its government.

[11] Interview: Michael Niejhar (Administrator), DGVII, Brussels, 20.1.97.

[12] Ibid.

[13] Ibid.

[14] AEA, 'Yearbook', 1996, p.32.

[15] ECAC and the Joint Aviation Authorities are regulatory, non policy making authorities (i.e. they are meant to police existing policy formulated by the Member States which comprise the bodies). However, the dominant states which comprise these bodies are, in reality, the Member States of the EU. ECAC consists of 33 states (including the 15 EU

members), and deals with the co-ordination of policy covering technical and economic aspects of air transport, facilitation and security for scheduled and non scheduled services. ECAC is an association, but does not have the power of enforcement. The JAA is the co-ordinating body for 23 states (including all EU members), but does not have the power of enforcement of standards. This power still resides at the national level, but many air transport interests wish to see this power integrated as an arm of the EU. For further information, see: 'An Unsafe Idea' (editorial comment), *Flight International* (10-16 Apr, 1996), p.3; 'Without Authority' (editorial comment), *Flight International* (20-26 Nov, 1996), p.3; 'European FAA?' (editorial comment), *Flight International* (24-30 Jan, 1996), p.3.

[16] AEA, 'Yearbook', 1996, p.32.

[17] This term was first used in the context of the US Congress. Its definition is that a legislator will vote in the favour of another legislator's policy in return for a complementary vote at a later date.

[18] Interview: AEA, Brussels, op.cit.

[19] Interview: Steve Aylward (Manager, Competition and Regulatory Affairs), British Airways, Heathrow, 6.9.96.

[20] See case study in Chapters 5 and 6.

[21] Interview: British Airways, op.cit.

[22] AEA, 'Yearbook', 1996, p.32.

[23] The number of seats operated by an ERA member averages out at 157 (average total per airline). This calculation is based on the average number of seats per aircraft of ERA members (55), multiplied by the average number of aircraft in each member's fleet (2.86). Figures obtained from ERA, 'Yearbook', 1996/97, p.51.

[24] This calculation is based on multiplying the number and type of aircraft operated by each airline by the number of seats each aircraft carries. Sources: AEA 'Yearbook', 1996, pp.38-62, and Jane's / Collins, *Civil Aircraft* (Glasgow: Harper Collins, 1996).

[25] e.g. Jonathan Bailey (Manager, Aeropolitical Affairs), Virgin Atlantic, argues this point. (Crawley, 9.4.97).

[26] Press conference to announce House of Commons Transport Select Committee Report on the proposed alliance between British Airways and American Airlines, London, 30.7.96. See also: D. Kyrou, 'Air Transport Policy in the European Union - State Control By Other Means?', paper presented at the Research Conference of the University Association for Contemporary European Studies, Loughborough University, September 1997.

[27] House of Commons, Transport Committee, Sixth Report, 'The Proposed Alliance between British Airways and American Airlines', 30.7.96.

[28] Interview: Tressa Clay (Air Transport Section), UK Department of Transport, London, 7.4.97.

[29] The 'consumer protection' arm of the UK Civil Aviation Authority.

[30] Interview: Dolores O'Reilly (Air Transport Users Council), European University Institute (via telephone), Florence, 30.4.97. See also: O'Reilly, 'Testing Integration Theories: The Development of EU Air Transport Policy', paper presented at the University Association for Contemporary European Studies, Loughborough University, 11.9.97.

[31] Ibid. State aid is the funding of an airline by its host state; for example France injecting capital into Air France (the airline Virgin Atlantic argues that state aid can be indirect; for example, if the UK government chooses to prioritise BA in its air transport policy). The EU legality or otherwise of this practice is outside the remit of this research. However,

its existence may demonstrate the close relationship between states and their airlines. For details see: Mark Clough (Ashurst Morris Crisp, London), 'The Regulation of state aid in Air Transport', paper presented at IBC conference on EU air transport liberalisation, Brussels, 7.2.97. See also: Trevor Soames and Alain Ryan (Norton Rose), 'State Aid and Air Transport', in *European Competition Law Review,* Vol.16, No.5 (Jul-Aug. 95),pp.290-309; Simon Bishop, 'The European Commission's Policy Towards State Aid - A Role for Rigorous Competitive Analysis', in *European Competition Law Review,* Vol.18, No.2 (Mar, 1997), pp.84-88; Gerrit Schohe, 'Rivals in a state', in *Airline Business,* Vol.12, No.8 (Aug, 1996), p.42-45.

[32] Interview: British Airways, Heathrow, op.cit.

[33] Ibid.

[34] Although, as some of these airlines might well claim, it would not be possible for them to operate without such financial assistance. The key test for the Commission, in deciding whether to grant permission for a Member State to grant state aid, is whether competition is adversely distorted by such aid. See references f/n 35.

[35] There exists a wide level of debate on the precise definition of a 'new entrant' According to Council Regulation 95/93 on slot-allocation, a new entrant is defined as a carrier holding less than 3% of the slots at a particular airport. Additionally, an airline must not hold more than four slots on any day (for example, two take-off slots and two landing slots). The slots can serve any particular route. Therefore, an airline wishing to establish a new route which does not fulfil this criteria is not categorised as a new entrant (see Article 2 (b), 'Council Regulation (EEC) No. 95/93 of 18 January 1993 on common rules for the allocation of slots at Community airports'. O.J. L 14/1, 22.1.93). However, there exists debate as to whether or not an airline is a new entrant at an airport if it fulfils the above criteria at a particular airport, yet acts as a 'franchise' or 'wet lease' partner for an airline which does not fulfil the criteria. For example, the UK airline British Regional Airlines acts as a franchise partner for British Airways at London Gatwick Airport. Although fulfilling the criteria of a new entrant, the airline operates on behalf of British Airways at the airport. There is therefore debate as to whether the airline would qualify for 'new entrant status' at this airport. (See: Evidence to House of Commons Transport Committee on Regional Air Services, Terry Liddiard (Managing Director), British Regional Airlines, London, 1.4.98.) A franchise airline is a carrier who pays to use the name and marketing of a larger, more dominant airline. The franchise will operate routes on behalf of the dominant carrier, - routes which the dominant carrier is not be able to operate. For further information on franchising and 'code sharing', see: Chris Allen (Head of Competition and Regulatory Affairs), British Airways, 'The Role of Code Sharing and Franchising in British Airways' Development'; Malcolm Hart (Managing Director), Gill Airways, 'Partnerships- Our Path to Success'; Mike Bathgate (Commercial Director), British Regional Airlines, paper on franchising. All three papers presented at Royal Aeronautical Society, 'Franchising, Code Sharing and Related Matters', conference, London, 11.2.98.

[36] It might be argued that employment in national champion firms is kept artificially high (both by the state and the firm), in order that the company retains its national champion / elite status.

[37] The majority of Sabena shares were held by the Belgian government until 1996. In that year it retained a 33.8% stake but sold its controlling interest to the airline Swissair.

[38] Although, in June 1999 the Finish government appointed the merchant bank Credit Suisse First Boston to advise on the possibility of selling its 58% stake in Finnair. See: *The Financial Times,* 14.6.99, p.32.

[39] From published material, only the Belgian regional airline 'Delta Air Transport' proved non profitable in 1996 (source: *Airline Business,* Vol.13, No.5 (May, 1997), pp.30-33). It must be noted however, that the source for this information details the profitability of only 32 of the largest 50 regional carriers world-wide. However, the point remains that, because of their private ownership, most regional airlines cannot depend on state support in the way that many of the flag carriers can.

[40] Interview, British Airways, op.cit.

[41] Interview: British Airways, Heathrow, op.cit.

[42] Following the completion of this research the European Regional Airlines' Association changed its name to the European Regions Airline Association. The acronym ERA remains.

[43] See membership costs, below.

[44] Interview: Ian Woodley (Chief Executive), Business Air, Ballasalla, Isle of Man, 9.4.97.

[45] ERA, 'Yearbook', 1995/96, p.15.

[46] For a discussion of the relative importance of regional carriers, see: AJ Reynolds-Feighan, 'European and American Approaches to Air Transport Liberalization - Some Implications for Small Communities', in *Transport Research, Policy and Practice,* Vol.29, No.6 (1995), pp.467-483.

[47] However, a higher turnover does not necessarily indicate profitability. See section on the AEA below.

[48] Interview: Malcolm Hart (Chief Executive Officer), Gill Airways, Newcastle, 19.3.97.

[49] Interview: Gill Airways, Newcastle, op.cit.

[50] Interview: Andrew Clarke (Air Transport Policy Manager), ERA, Chobham, Surrey, 23.1.97.

[51] Note Chapter 1.

[52] See Chapter 5.

[53] Interview: Ian Woodley (Director), British Regional Airlines (formally of Business Air), Ballasalla, Isle of Man, 9.4.97.

[54] Interview: ERA, Chobham, op.cit.

[55] The Joint Committee for Civil Aviation is a discussion forum operated by DGV (Employment, Industrial Relations and Social Affairs), of the Commission. It must be noted however that DGV plays a negligible role in the formulation of air transport policy. It does not Propose policy in the area (this is the task of DGVII), or act as an enforcer of competition in a *quasi-judicial* manner (this is the task of DGIV).
It is notable that non-elites focus a great deal of attention on matters of a technical nature. Elites put equal or greater importance on their relations with the policy maker. It may be noted that the former concentrate their energies on the policy, whilst the latter focus equal weight on the politics. For a discussion of the terms 'policy' and 'politics', see: Walter Kickert, 'The development in the study of public administration in Europe', paper presented at the Public Administration Committee, Chartered Institute of Public Finance and Accounting, London, 17.1.97.

[56] Interview: ERA, Chobham, op.cit.

[57] See references in Chapter 2, f/n No.2 for details.

[58] i.e. one that they do not perceive as a 'substantial' interest, such as a flag carrier.

[59] Interview: ERA, Chobham, op.cit.
[60] See Chapter 2, f/n No.2.
[61] ERA, 'Yearbook', 1995/96, p.14.
[62] Ibid, pp.19-94.
[63] Interview: Ian Woodley (Chief Executive), Business Air, Aberdeen, 17.5.96.
[64] Interview: ERA, Chobham, op.cit.
[65] AEA, 'Yearbook, 1996, pp.75-76.
[66] See: ERA, *Regional Report* (Jun, 1996), p.1.
[67] David Quick, 'Changing the brief', ERA, 'Yearbook', 1996/97, p.20.
It may be ascertained from this wording that ERA began to shift toward a forecasting role, as opposed to a responding role.
[68] ERA, 'Yearbook', 1996/97, p.27.
[69] Ibid, p.33.
The Operations Director of an airline, whilst being a technical expert, may well be unaware of the political ramifications of a situation, and with intricacies of EC policy formulation. However, most regional airlines rely on him / her in pursuing their position at ERA.
[70] Interview: ERA, Chobham, op.cit.
[71] See Chapter 5.
[72] This classification helps to define a clear division between subsidiary and non subsidiary airline members of ERA. However, many regional airlines have agreements with the flag carriers to co-operate in many different areas of business; thus there is an element of 'dependency' borne by non-subsidiary carriers, despite there being no ownership control by the flag carrier. Many non-subsidiary regional airlines require the co-operation of flag carriers to share terminal facilities, and several airlines, not featured in the table above are 'franchise' airlines of the flag carriers, who rely on the renewal of such agreements, by the flag carriers, in order to remain operationally viable. For further details see: Frances Butler-Sloss (Partner, Harbottle and Lewis), 'Airline franchising: Some EC law aspects', in, P.D.Dagtoglou, J.M.Balfour and S.A.Mirmina (eds), *European Air Law Association, Conference Papers, 9* (London: Kluwer Law International, 1996), pp.45-57.
[73] Interview: ERA member airline.
[74] KLM increased its share holding to 100% in July 1997; the above sources show the KLM share holding of Air UK as 45%.
[75] BA Regional is a subsidiary of BA. It is entirely separate from British Regional Airlines which is a non subsidiary regional airline.
[76] Acquired by British Airways, 1997.
[77] The airport operator may own several airports. For example, BAA (formerly the British Airports' Authority), is a member of ACI, yet collectively owns seven airports which it subsequently represents.
[78] With the possible exception of ADV (examined below), whose membership includes Austrian and Swiss airports.
[79] See above.
[80] Calculation from ACI Europe documentation.
[81] ACI Europe.
[82] Again note Jordan's comments on interest group 'legitimisation' (see Chapter 1). Indeed, it can be argued that the legitimacy of ACI is substantially increased by the fact that it is a

non exclusionary group, as opposed to an 'elite' group which allows only two non-elite members, as is the situation in the AEA.

83 Airports that are not members broadly fit into two categories. A large number of airports in the former Soviet Union have not yet applied for membership. Additionally, a number of very small airports (in terms of passenger numbers), serving domestic destinations are not members (for example, in the UK, airports such as Stornoway and Kirkwall in Scotland).

84 ACI, 'ACI, European Region, English Edition'.

85 See below.

86 A 'traffic unit' provides a tool to compare airports. Each unit (as devised by ACI), consists of one passenger, or 100 kilograms of cargo. However, it does not include transit (ongoing passengers), as a unit. This figure is not insignificant; for example, at Heathrow (UK's largest airport), one third of all passengers are in transit.

87 The majority of the other 21 airports in the UK are owned by local county councils. In Member States where the regional authority possesses a great deal of power over the national executive (such as Germany), it may be the case that the regional authority (for example, the Länder), of a particular state can influence the formulation of relevant EC policy.

88 The members consulted and involved in the writing of and thus represented by the document are as follows (a figure in brackets denotes the number of airports represented by the member, if more than one): Dusseldorf, Germany; Vienna, Austria; BAA, UK (7); Malta, Aer Rianta, Ireland (3); Frankfurt, Germany; Rome, Italy; Aeroports de Paris, France (3); Copenhagen Airports, Denmark (2), Zurich Airport Authority, Switzerland; Amsterdam, Netherlands; Madrid, Spain; Arlanda, Sweden (19); ANA, Portugal (7), Milan, Italy. See: ACI Europe, 'Consultations on ground-handling services. Response by ACI Europe to the European Commission, DGIV and DGVII.' 28.4.94.

89 Members of the Board are elected at the Regional Assembly, held annually.

90 Interview: ACI Europe, Brussels, op.cit. The other committees are: Communications, Economics, Environment, Facilitation and Security, Human Resources, Legal Affairs, Marketing and Commercial Strategy, Technical and Safety.

91 Interview: Shenna Gibson (Head of European Policy Unit), ACI Europe, Brussels, 10.4.97.

92 Interview: ACI Europe, Brussels, op.cit.

93 See below.

94 Interview: Bervd Nierobisch (Executive Director), ADV, Stuttgart, 10.7.97.

95 Interview: Alastair McDermid (Head of Airport Strategy), BAA, London, 22.7.97.

96 Interview: Frank Pullman (Chief Executive), Luton Airport, Luton, 6.1.97.
As a 'snap shot' comparison with BAA (annual traffic of 101 217 908) the annual traffic unit figure for Luton is 1 192 951 (Source: ACI Europe).

97 Interview: Luton Airport, Luton, op.cit.

98 Interview: Reg Bench (Air Transport Solicitor, Williamson and Horrocks), London, 16.1.97.

99 Additionally, ADV also represents Austrian airports and Zurich airport (as Zurich is located within the Germanic speaking section of Switzerland).

100 Although airport members of these groups are also individually represented within ACI.

[101] The exceptions to this rule, the states which have a single airport operator for the entire country, are Ireland, Spain and Finland. The latter was not an EU member at the time of the formulation of the ground-handling Proposal.
[102] Interview: BAA, London, op.cit.
[103] Interview: Richard Gooding (Chief Executive), London City Airport, 9.4.97.
[104] Airports in this category include: Hamburg, Frankfurt, Munich, Cologne and Berlin.
[105] Airports in this second category include: Stuttgart, Nuremburg, Hanover and Düsseldorf.
[106] Interview: ADV, Stuttgart, op.cit.
[107] Ibid.
[108] Ibid.
[109] Note the study by Jordan *et al* of the representation of automobile industry (by the AEAC), in Chapter 2. The group substantially improved its interest representation when decisions were made by majority, rather than unanimous voting. See, Andrew M.McLaughlin, Grant Jordan and William A. Maloney, 'Corporate Lobbying in the European Community', in *Journal of Common Market Studies,* Vol.31, No.2 (June 1993), pp.191-211.
[110] The number of votes a member carries within ADV depends on the throughput of passengers at that particular airport; Frankfurt has the largest single number of votes.
[111] Note Greenwood, in Chapter 1. Greenwood stresses that effective interest representation requires the larger interests dominating a group's position. Formalising QMV into the constitution of ADV ensures that this happens.
[112] Interview: ADV, Stuttgart, op.cit.
[113] Ibid.
[114] The decision making and decision implementation reflects the examination of the AEAC examined in Chapter 1.
[115] Interview: BAA, London, op.cit.
[116] See Chapter 5.
[117] Ibid.
[118] Interview: ADV, Stuttgart, op.cit.
[119] Ibid.
[120] See Chapter 3. See also: Dolores O'Reilly, 'From State-Centric Control to EC Competence', paper presented at European University Institute, Florence, 19.3.97.
[121]One of the few exceptions to this rule was the independent scheduled airline British Midland, which joined the AEA in 1993, forty years after establishing itself; British Midland was refused admission until this time (see: Simon Bulmer, 'Air Transport liberalisation (unpublished manuscript), University of Manchester, 1997, p.17). In 1984 Ryan Air, a 'low cost' scheduled carrier began services from Dublin, n 1995 the airline Easyjet was established in the UK, and in 1996 the airline Citybird was established in Belgium. Like Virgin all three are independent scheduled carriers which are not members of AEA or ERA.
[122] A charter airline is any airline that does not publish its flight details in the 'ABC Directory' an international directory (in both paper and on-line computer format available to booking agents), published by IATA (the International Air Transport Association). The defining characteristic of a charter airline is that such an airline rents out the aircraft to a company or companies that then sells the seats to a consumer, thus lowering the price significantly. The passenger is highly cost conscious, and is likely to use the flight for leisure, rather than business, because of the infrequency, and lack of flexibility of the

service (that is the ability of the passenger to alter the time or destination of their flight once booked).

[123] Senior official within Virgin Atlantic.

[124] Interview: Dr. Barry Humphreys, Director of Government and External Affairs, Virgin Atlantic, Crawley, 3.9.96.

[125] Ibid. Although this advantage is offset by the fact that flag carrier airlines (AEA members), such as BA, enjoy privileged, non official (*id est, ad hoc*) contact with officials in the Commission. Nevertheless, this factor would place Virgin at an advantage to ERA members which may not enjoy such non official contact.

[126] Although, prior to 1996 Virgin had no Department of External Affairs. In that year Dr Barry Humphreys joined Virgin from his previous position as Head of Air Services Policy and Industry Affairs at the UK Civil Aviation Authority. In 1997, a Manager of Aeropolitical Affairs was appointed. (This individual, Jonathan Bailey, was working in a senior position within BA when he was appointed to his new role at Virgin!) Prior to 1997 (and thus the formulation of the ground-handling Proposal), Virgin played an *ad hoc* role in such interest representation.

[127] 'Servisair PLC ,'Annual Reports and Accounts' 1996 (calculation from), p.1.

[128] The task of ground-handling is separated into 'land-side handling' and 'air-side handling'. Definitions of the two are provided in Chapter 5; however, it is the latter of the two that provided the contention surrounding the formulation of the Proposal. A full service encompasses both passenger and ramp handling. Additionally, Servisair provides 'supervision' (where a Servisair receive a fee for monitoring and controlling other company's ground-handling services), at a number of EU airports.

[129] Servisair's financial turnover, 1996, £147.1m (source: Servisair PLC, 'Annual Reports and Accounts' 1996, p.18). As a comparison, the largest EU airline, British Airways has a turnover of over £7 500m whilst the largest non subsidiary regional airline in the EU, Eurowings, has a turnover of over $305m (£191m) (sources: *Flight International,* 'The Top Fifty Airlines' (31 Jul,- 6 Aug, 1996), pp.31-38 and *Airline Business,* Vol.13, No. 5 (May), 1997, p.30).

[130] Interview: Anthony Rowbotham (General Manager, Business Development), Servisair, Manchester, 16.7.97.

[131] Belgavia also own several subsidiary ground-handling operations in France.

[132] Interview: Servisair, Manchester, op.cit.

[133] See Chapter 1.

[134] Interview: BAA, London, op.cit. The exception to this is Zurich Airport. As a member of ADV, yet a non EU member airport, it can be argued that this airport possesses more influence on EC policy formulation than any other EU or non-EU member.

[135] States which have prospects of membership of the EU in the near future. During the formulation of the ground-handling Proposal, Vienna Airport, Malta Airport and the main Swedish operators were actively involved. (Sources: ACI, 'Consultations...', op.cit.)

[136] The exception to this rule is Cross Air (a subsidiary of Swiss Air), although Swiss Air have a controlling interest in the Belgian national flag carrier, Sabena.

[137] Note the example of British Midland, above.

[138] Although this does not necessarily enhance the decision making procedure of the AEA or enhance its lobbying capabilities.

[139] Interview: A senior member of the Transport and Tourism Committee of the European Parliament, Brussels, 20.4.98.

[140] Source of significant authority, formerly at BA.
[141] Interview: BAA, London, op.cit.
[142] See Chapter 3, f/n Nos. 53, 54 and 56.
[143] Interview: Gill Airways, Newcastle, op.cit.
[144] Interview: ADV, Stuttgart, op.cit.
[145] Interview: London City Airport, London, op.cit.

5 The Positions of the Interest Groups

The case studies contained within Chapter 6 focus upon the issues of contention between the interest groups the Association of European Airlines, the European Regional Airlines Association, and Airports Council International. Having examined the composition of these groups in the previous chapter, this chapter will examine the positions taken by the groups in these case studies of ground-handling and slot-allocation.

Ground-handling

Ground-handling is an essential service required by airlines at all the airports which they operate from. The relative cost of this service to airlines has a considerable bearing on their economic efficiency. However, this cost varies considerably, dependent upon the airport which is being operated.[1]

Ground-handling can broadly be separated into two main areas - landside and airside-handling.[2] Landside-handling, '...comprises any kind of assistance to arriving, departing or transfer passengers, in or outside the airport, and includes passenger check in'.[3] Airside-handling is essentially all other services required by an airline. These services include:

- Baggage handling (to / from the aircraft, including the sorting of baggage in the terminal building).
- Freight and mail handling.
- 'Ramp' handling (towing, parking, ground communication, loading / unloading aircraft).
- Provision of fuel / oil for the aircraft.
- Safety provisions.
- Catering.
- Aircraft servicing / maintenance.[4]

Both airside and landside-handling functions are operations which are equally important for the operation of an airline.

Landside-handling (otherwise defined as passenger handling), is particularly important for the 'image' of an airline. Airlines often wish to employ their own personnel for this operation to ensure the best customer contact possible; airlines are very often reluctant to allow personnel from other airlines to passenger handle as they may not offer a service that meets their particular expectations. However, in the formulation of the ground-handling Proposal, landside liberalisation was largely a non-contentious issue; it was the liberalisation of airside-handling which presented itself as the main issue of contention in the formulation of the Proposal for legislation. The key reason for this is explained by an executive of the ground-handling firm Servisair:'...there's not a great deal of money in passenger handling, the key is airside.....that is where the money is and where the major concerns are'.[5]

Definitions of 'more' and 'less' liberal ground-handling markets

Ground-handling services can be delivered in a combination of ways.

• The airport can deliver the ground-handling services. For example, Frankfurt Airport can provide passenger handling, re-fuelling and the other services to airlines such as Lufthansa and British Airways required for the operation of their aircraft operating from this airport.
• The home flag carrier can provide a monopoly handling service. For example, Olympic Airways were the sole supplier of ground-handling services at Athens airport (rather than the airport itself).
• An airline can 'self-handle'. This is where some or all of the above services are carried out by the airline that wishes to conduct these services. For example, Aer Lingus operates the equipment and employs the personnel to re-fuel their aircraft and execute the other ground-handling services it requires at London Heathrow Airport.
• An airline can hire a 'third-party' to handle one or more of the above services. For example, British Regional Airlines at London Stansted Airport contract the firm Servisair for its ground-handling activities. Normally, the greater the number of such third-party handlers, the greater the choice and the greater the cost competitiveness.

At 39% of EU airports there existed a monopoly on the landside,[6] although the situation regarding landside liberalisation has been improving steadily since the formulation of the ground-handling Proposal in 1996.[7] The main issue of contention was the fact that 96% of airports supplied a monopoly on the airside.[8] Several airlines alleged that such a situation created unnecessarily high costs, resulting from their inability either to self-handle or to exercise a choice between third-party ground-handlers. This 'more liberal' camp included the airlines that wished to exercise a choice of third-party handlers, as well as those that wished to self handle. The 'less' liberal camp included the airlines that wished to restrict (either overtly or covertly), the choice of third-party handlers and the ability of other airlines to self-handle. This position was usually established because certain airlines wished to protect their own ground-handling markets. As examined below, the groups the Association of European Airlines (AEA), the European Regional Airlines Association (ERA), and Airports Council International (ACI), can broadly be divided into those that wanted a 'more', and those that wanted a 'less' liberal market in the provision of ground-handling services; ACI appeared to be oppose liberalisation, ERA firmly supported liberalisation, whilst the AEA formulated a somewhat weak, 'diluted' position on the issue as a result of disagreement within the group.

The position of the Association of European Airlines

As stated in the previous chapter, the AEA is a group consisting of elite national champion interests. However, the member airline British Airways has noted the 'increasing gulf' between state owned (normally non-profitable) airlines and non state owned (profitable) airlines within the AEA.[9] Although the formal decision making procedure of the AEA allows majority (hence firmer) decision making, in practice positions of the AEA are established by unanimity; this was the situation the case study of ground-handling. Broadly, the AEA established a liberal view on the issue. However, despite the eliteness of its membership composition, the AEA established by far the weakest position of the three interest groups examined. This weak position arguably derived from its *de facto* method of position formulation. Indeed, an AEA member stated: '...the AEA failed to form a position on ground-handling until it was far too late'.[10]

Unlike the ERA, and (in particular), ACI who both presented highly detailed proposals on the format of the liberalisation that they required, the AEA's proposals were considerably more 'general', consisting only of broad objectives.

Airlines are customers of ground-handling services. As customers, one might have expected that they would require the most liberal market available in which to purchase these services. One might also imagine that they would wish to have the ability to self-handle if they so choose. However, airlines are also providers of ground-handling services. In several EU airports the flag carrier airline is the sole (hence the monopoly), provider of these services. Such carriers felt that the ground-handling market, in which they had, and continue to have, a monopoly, and the exclusivity of access should be maintained for themselves.

The original complaints concerning the issue of the monopolisation of ground-handling services by airports and airlines were put forward by British Airways, KLM and SAS to DGIV (Competition), of the European Commission.[11] These airlines are privately owned, profitable carriers (albeit national champion, 'elite' firms). These airlines wished to have the ability to self-handle (if they so chose), and the ability to choose from a wide range of third-party handlers. However, the state owned AEA members Iberia, Olympic and TAP supplied monopoly ground-handling services in Spain, Greece and Portugal respectively.[12] Iberia, Olympic and TAP vociferously opposed liberalisation and therefore opposed the objectives of the above airlines. Subsequently, '...there was a conflict of interest as some of the flag carriers were part of the monopoly ground-handling situation'.[13]

The situation in France and Italy was somewhat different. In these Member States there existed a monopoly ground-handling situation. However, the monopoly was provided by the airports rather than the airlines. Furthermore, the owners of these airports - the governments - are also the owners of the AEA member airlines. These airlines were in a position whereby they themselves would benefit from a liberalised market as their ground-handling costs would, in all probability decrease, as a result of the increased competition. However, their owners (i.e. their governments), were also the suppliers of ground-handling services and wished to preserve their monopolies at the airports. The airlines Air France and Alitalia were in a position where the directors of these airlines seemed to oppose the positions of their own governments. There therefore existed a situation in France and Italy where the elite (state owned) airline was pitted against the elite (state owned) airports.

In France the elite airport interest, *Airports de Paris* may appear to the casual observer to have a liberalised ground-handling market, as there are three ground-handlers to choose from at each of the airports under its control. However, all three ground-handling companies are owned by the French state, and some quarters quite rightly argued that such ownership restricts competition.[14] Others (in particular ACI), argue that, despite the fact that an airport or its ground-handling firms are owned by a single state or company, there is not a sufficient 'link' between the state and the ground-handling firms to stifle competition.[15] Air France would clearly benefit from ground-handling liberalisation *per se*, and Air France and Alitalia were signatories to the original complaint to DGIV. A test was therefore created between the 'eliteness' of the airlines and the airports in question. Despite the close relationship that existed, and continues to exist between these airlines and their governments, the governments sided with the airports. The positions of Air France and Alitalia on ground-handling may have reflected the views of the directors of these firms, but they did not, on the whole, reflect the views of the majority of employees or their unions. In the case of Air France, this was a contributory factor which ultimately led to the dismissal of its Chairman in November 1997.[16]

The German flag carrier, Lufthansa, presents a different situation. It is majority privately owned, although the German government and public institutions continue to own 39.65% of the airline. Moreover, the airports in Germany (all of which are publicly owned), provided monopoly ground-handling services. The position of the German airports group ADV (examined below), and the German state were the least liberal of all the views put forward. However, Lufthansa was also a signatory to the original complaint to DGIV. Like Air France and Alitalia, Lufthansa, as an elite airline, was pitted against the elite airport interest in its own particular state.[17] A crucial difference between these cases is that the German government had no power to dismiss the Chairman or other directors of Lufthansa (despite the fact that the airline is a flag carrier), as the government and public institutions possess only a minority stake in Lufthansa.

BA, KLM and SAS were, however, the most vociferous supporters of the ability to self-handle and to have a wide choice of third-party handlers. Uniquely, these three airlines had the support of their governments.[18] These airlines were also supported in their stance by British Midland upon their membership of the AEA in 1993.[19]

Within the AEA there was a compromise in order to appease the Portuguese, Spanish and Greek interests. Circumstances indicate that it was primarily these 'political dimensions'[20] *within* the AEA which were of the most concern in producing its position.

> ...interests were diverse, so it took a longer time (sic) to work out a clear position...when there are competitive interests which are diametrically opposed in any group it became difficult to achieve a joint statement.[21]

The Compromise Position

The AEA initially formulated a 'briefing paper' for the Commission. However, a fundamental problem of the briefing paper, according to the AEA itself, was that it was prepared by 'experts in the field' and, 'these experts were unaware of the political dimensions of the process of legislation at Community level'.[22]

The AEA claimed to support a liberalised ground-handling market. As stated by British Airways, individual airlines approached the Commission to make their case, using their ability to lobby at Level III (directly to Brussels). The position of the AEA was a lowest common denominator position[23] as a result of its decision making structure being based on a unanimous consensus.

> Some of the airlines are also the monopoly handlers and they refused to go along with the view that that the rest of the airlines wanted to take. There were three airlines which refused to go along with what the rest of us requested.. Individually we all...requested [what we wanted] as the AEA position was diluted.[24]

The position of the AEA does, however, more closely resemble the view of the liberal camp (BA *et al*), than that of the 'non-liberal' camp (TAP *et al*), although there was a significant compromise by the liberalising group.

The AEA supported the use by DGIV of a Commission Article 90 Directive which would have been aimed at the airports about which the original complaints were made.[25] An Article 90 investigation and subsequent Commission Directive would have dealt with the complainants and set a precedent for EU airports. An official from the AEA stated:

> ...the AEA airlines regret that Article 90 was not used. This single instrument should have been used arising from the complaints in one or two key cases in order to arrive at key legal decisions.[26]

Nonetheless, the eventual position, decided by a unanimous consensus, stated:

> A greater degree of choice for airlines to choose their handling agencies would evidently bring the benefits of competition to this important sector. To that effect, the European Commission has developed a proposed Directive aimed at widening access to the market, while taking account of the constraints.
>
> However, the substance of the [proposed] Directive falls short of its aims; on the landside, a two-year transition period is unnecessarily long for this less complex matter. On the ramp [airside], it allows Member States to invoke physical constraints in limiting the choice to two suppliers, while not imposing any obligation on them to justify their decisions on the basis of objective criteria.[27]

This statement was broadly a pro-liberal one, emphasising the need to expand the choice of third-party handlers, and querying the possibility that Member States may be able to use the criteria in order to circumnavigate ground-handling liberalisation. However, in order to appease TAP *et al*, the statement made clear that there were 'constraints' which might limit the feasibility of ground-handling liberalisation.

The initial pro-liberalising airlines (BA/KLM/SAS) wished to insert an amendment to the position of the AEA airlines on the matter of 'self-handling'. The definition of self-handling in the eventual Proposal is one where an airline is permitted to provide handling services for itself or a subsidiary. However, it is criticised by BA/KLM/SAS as being unduly restrictive as it does not take into account various types of alliances, such as franchising.[28] For example, if the ability to self-handle was enforced at airports across the EU, BA would be able to self-handle at Brussels Airport, if it chose to do so. BA would also be able to handle for 'British Airways

Regional' as BA Regional is a subsidiary of BA. However, the airline British Regional Airlines is a franchise operation of British Airways. BA do not own a majority of the shares of British Regional Airlines and, therefore, British Airways would not have the automatic right to handle British Regional Airlines' aircraft at Brussels Airport even if self-handling was permitted. BA/KLM/SAS therefore wished to formulate an AEA position requesting that the definition of self-handling be 'broad' as opposed to 'narrow'. However, the majority of AEA airlines opposed this;[29] this included the 'monopolistic airlines' (TAP *et al*), in addition to the other signatories to the original complaint - Air France, Alitalia and Lufthansa. It was subsequently not included in the final AEA position, although BA/KLM/SAS attempted to pursue this position independently.[30] In sum, it is arguable that the AEA produced a weak, Lowest Common Denominator position on the issue of ground-handling as it was forced to produce the opinion by a unanimous consensus.

The Position of the European Regional Airlines' Association

Third-party and self-handling

ERA's position was considerably more detailed and precise than the position of the AEA. Nonetheless, ERA did not produce an official, public position on the matter until February 1995, two months after the publication of the Commission Proposal.[31]

In the introduction to its position, ERA highlights that ground-handling costs are 11% of the expenditure of an average regional airline. ERA's position is clear and not a lowest common denominator position.

> Carriers should have the right to handle their own services [i.e. self-handling]. Third-party handling should be permitted with either no, or minimal, limitation on the number of suppliers at individual airports. A self-handling carrier must also be able to offer third-party handling to other airlines...[32]

This position was an absolute. Unlike the AEA position there was no caveat which might allow Member States to impose restraints on the ability to self-handle, or impose a limitation on the number of third-party handlers that might be permitted.

The Commission Proposal for a Council Directive allowed a Member State to limit the number of third-party handlers to a minimum of two for various reasons concerning matters such as 'space, safety and technical feasibility'.[33] Additionally, the Proposed Directive allowed the Member State to limit the number of airlines that could self-handle for these same reasons.[34] However, ERA specifically stated,

> ...such a decision should be open to challenge and appeal. The European Commission and the airport Users Committee should be able to challenge such a decision with the Commission having the right to override it where it feels that the restriction is too limiting or unnecessary.[35]

Therefore, ERA stated that where limitations (to self-handling or the number of third-party handlers) might exist, due to space and / or capacity constraints, the users (airlines), should have formal access to the Commission to challenge and appeal against the decision. Furthermore, the Commission should be obliged to respond 'expeditiously' to such challenges. ERA also demanded that the right of appeal by airlines to the decisions of the Member States should apply to all aspects of the Directive; i.e., any decision by an airport under the auspices of an 'exemplary clause' (such as the right to limit self-handling), should be open to an appeals procedure.

'Equality of competition'

ERA stated that it was not sufficient to have a choice of ground-handlers; the choice must also ensure that there was fair competition between them.[36] ERA stated that regional airlines were often 'discriminated' against by ground-handlers, hence the increased need for a greater choice in the selection of ground-handlers by regional airlines. ERA cited the example that ground-handlers would often position regional aircraft a substantial distance from the airport terminal, resulting in the regional airline having to pay for the ground transportation of passengers from the aircraft to the terminal building, as well as resulting in inconvenience to the passengers. Another example of this ability of ground-handlers to 'favour' certain airlines is that often ground-handlers will allow the flag carrier airline at a particular airport to park their aircraft closer to the terminal building on the walk-way than other airlines.[37] True competition would therefore need to ensure that all ground-handlers possess the ability to compete on equal terms without having business opportunities 'stacked against them' in such a way.

'Minimum airport size'

This was a key issue for ERA. The proposed liberalisation of the ground-handling market applied only to airports whose annual traffic was not greater than a certain number of passengers.[38] According to Article 2 of the proposed Directive, airports whose annual traffic was less than two million passengers per annum were excluded from the criteria and therefore could continue to deny self-handling and prevent a choice of third-party ground-handlers *ad infinitum* until their annual traffic grew beyond this figure. ERA opposed any such minimum criterion, primarily because many of the airports its members operated from were airports below this threshold, and therefore its members would not be able to benefit from much of the ground-handling liberalisation. However, ERA attempted to be pragmatic about the issue and stated, '...if a limitation is to be imposed at all, then a threshold of one MPPA [million passengers per annum] is more relevant'.[39]

The airport 'Users Committee'

Article 5 of the Proposed Directive stated that each airport in the EU, above the 'minimum size' criterion should have a Users Committee[40] that consisted of all airlines using the airport to represent the interests of its users. However, the proposed Directive stated that the decision making procedure of this Committee '...may take account of the volume of activity of the various users of the airport in question...'[41] Such a decision making procedure would, according to ERA, have been of a substantial disadvantage to ERA members who by their nature carry a lesser volume of traffic in comparison to flag carriers. ERA therefore stated,

> ...the decision making process within the Committee should not be directly proportional to activity, otherwise smaller and new entrant carriers could be severely disadvantaged.[42]

Therefore ERA was opposed to any decision making procedure of the Committee that gave an advantage to the larger, elite carriers.

In summary, despite the fact that ERA is a non-elite group (and arguably weaker in terms of its interest representation), its position on ground-handling was concentrated in comparison to the diluted position of the AEA.

The position of Airports Council International

Ground-handling is an issue which presents a near unique situation to the group that represents airports. The imposition of a minimum level of ground-handling competition would affect all airlines in the EU. However, if a particular airport has a restricted level of ground-handling competition this situation would not affect (adversely) other airports with a more liberal level of competition. Therefore, there is no incentive for members of a group such as ACI (whether they are elites or non-elites), to oppose those interests which demand that the group took a 'minimalist', less liberal view. For example, at Heathrow Airport, operated by BAA, there is a choice of eight ground-handlers; it is the most liberal airport in the EU, and BAA sees operating a liberal ground-handling market as the surest way to maximise its own profits. Consequently, two thirds of the profits of BAA are derived from the retail sector, whilst the bulk of the remainder is derived from landing charges; a liberal ground-handling market would therefore not adversely affect BAA. However, the view of ADV (the German airports' group), was that the EU should regulate for a less liberal environment, as a substantial proportion of the revenue to its airports derives from ground-handling activities.[43] If ACI were to take a view that a Directive should, for example, feature a statutory minimum number of ground-handlers (i.e. a less liberal view), this position would not effect those that were already 'more liberal'. Such airports could continue to operate a liberal ground-handling market if they so chose. This hypothesis is supported by the airlines group, the AEA:

> ...the AEA interests were diverse, on this issue. British Airports were fairly liberalised prior to the Directive. Airports do not move around, unlike airlines, therefore the position of the German airports does not really matter to the British airports [i.e. BAA]. You could say that the liberalised airports would not really care about the non liberalised

> airports. Therefore there is no urgent necessity for them to agree. This means that [it is] far easier for a group of airports within an association such as ACI to take the lead under the banner of the association and promote their own interests than it is for the airlines to do so.[44]

The debate within ACI therefore concerned the minimum level of ground-handling liberalisation to be requested by the interest group. Those with the least liberal ground-handling markets were interested in keeping this level as low as possible. Those with more liberal markets were not overly concerned as they could retain their liberal ground-handling situations.[45] Like the Association of European Airlines, ACI therefore produced a lowest common denominator position on ground-handling. However, this lowest common denominator position was one which did not adversely affect those interests who would have been content to have produced a 'more liberal' position. It was not, therefore, an LCD reached through hard bargaining. It was a position that interests such as BAA, although academically opposed to, did not oppose within ACI as it would not have benefited BAA to have done so. This is in contrast to the AEA, where the LCD position produced adversely affected the members that wished to take a more liberal position.

From the three interest groups examined, the earliest position established (as well as the most effective position), was that of ACI. It established various official positions both prior to and following the formulation of the Commission Proposal on ground-handling. Official and public positions were established in 1991,[46] 1992[47] (prior to the Commission's announcement of the original complaints concerning ground-handling),[48] and 1993.[49] Furthermore, in 1994[50] prior to the formulation of the Commission Proposal, a comprehensive document was produced for DGIV (Competition), and DGVII (Transport). This was followed by an ACI Resolution in the same year.[51] Only one official position was adopted after the Commission published its original Proposal.[52] Therefore, ACI was not only the first to forecast the legislation but was also the key informational resource for the Commission when drafting its Proposal, particularly via its comprehensive 1994 document. Furthermore, all of these documents were formulated primarily with the Commission (both DGs IV and VII), in mind. The Commission was also the first body to see the documents prior to their official publication.[53] Neither the AEA or ERA had provided such a comprehensive source of information prior to the formulation of the Proposal.[54]

ACI Europe firmly opposed the use by DGIV of a Commission Article 90 Directive.[55] Such a Directive would have targeted the airports about which the original complaints were made by British Airways, KLM and SAS.[56] As stated above, such a Directive would have set a precedent which would have forced monopolistic airports to liberalise. Indeed, when the Competition Commissioner (Van Miert), announced the airline complaints in August 1993, he stated his preference for a Commission Article 90 Directive.[57] However, two months before this announcement, ACI stated:

> ...any regulatory regime in this area must apply equally to airlines, airports and independent handling agencies in order to safeguard the interests of airport users.[58]

All the positions adopted by ACI repeatedly point to its belief that:

> ...the airport should decide the degree to which independent or airline handlers, self-handling and third-party handling should be permitted at their particular location [and] the only organisation capable of striking the right balance between the objectives of efficient provision of ground-handling, safety and security whilst minimising cost] is the airport itself.[59]

This view was stated back in 1991, and repeated a year later: 'The airport has to determine the way in which the handling services are performed...'[60] In 1993 ACI also stated:

> ...it is essential in order to safeguard the interests of passengers, airlines and all airport users, for airport operators to have the prime responsibility for determining ground-handling policy at their airports.[61]

Probably the clearest declaration against the use of an Article 90 Directive by the Commission can be found in the February 1994 document. ACI stated:

> Should a directive be adopted by the European Commission,[62] without political control[63] it would, in the present circumstances, clearly be an arbitrary act and consequently an illegal misuse of power.[64]

adding:

> ACI Europe must therefore conclude by respectfully inviting the European Commission to abandon its notion of intervening in airport ground-handling by means of a directive, especially one based on Article 90 of the Treaty.[65]

ACI therefore had a history of consistency in its views against the use of a Directive by DGIV in matters of airport ground-handling.[66] This was expressed implicitly prior to DGIV's announcement of complaints, and then explicitly when it seemed very possible that DGIV would produce such a Directive. The February 1994 document made clear ACI's prime objectives.

Unlike ERA, ACI stated that the ownership of an airport is irrelevant in terms of competition, as the management is kept separate from the ownership. Subsequently, in the situation of *Airports de Paris*, where three seemingly 'independent' ground-handlers are all state owned, ACI would seem to claim that the three airports were *de facto* separate from one another (i.e. no monopoly exists in such a situation),[67] although to the impartial observer, that would be difficult to believe.

Third-party handling

ACI objected to any legislation which would stipulate a minimum number of ground-handlers that must be available at an airport on either the air, or land side. ACI stated three main reasons for this position.

1. A minimum criterion would infringe on an airport's right of management control. Airport operators have a statutory responsibility for the 'safe and efficient management of their airports'.[68]

2. It is a 'contradiction in terms' that to increase competition, the market must be regulated to ensure that there are a minimum number of suppliers which the market may not demand.[69]
3. Setting a minimum number of ground-handlers would potentially, or actually, interfere with the contractual rights of those already supplying ground-handling services, if an Article 90 Directive was to be used to enforce this minimum number.[70]

Self handling and third-party handling

ACI's February 1994 consultation paper stated:

> ACI Europe advocates discarding the notion of self-handling altogether, whilst making it possible for all airlines to bid for airside ground-handling contracts in the same way as independent ground-handling firms.[71]

i.e. if airlines wish to third-party handle they must apply to the airport operator who may accept, reject or alter the application. There must be no 'right' of airlines to self-handle on the airside, and the choice must be left to the individual airport operator.

ACI claimed that it established these positions after consultations with fifteen European Airports.[72] At most of these airports it was claimed that 'special situations' existed that would prevent the liberalisation (or further liberalisation) of ground-handling services. The participants in these consultations represented the airports (or their operators), that were the largest airports in each of the EU members states (other than Greece, Luxembourg and Belgium). Thus, with the exception of Athens, all the airports, about which the original complaints were made, were featured in the consultation exercise.

The reasons for the rejection by ACI of airlines' ability to self-handle, can be summarised as issues of safety, space constraints, security, technical feasibility and additional costs that would have to be met by the airports.

In terms of safety, ACI stated that the introduction of more ground-handlers on the airside would result in an increase in personnel, equipment, and vehicles, and hence an increased risk of accidents. The airport operator must therefore be able to implement and monitor safety requirements which may result in limiting the number of ground-handlers. Lack of space to facilitate further ground-handlers was a problem at many airports, especially in the provision of the storage of equipment and vehicles, ACI claimed. Attempts to increase available space might result in the reduction of aircraft capacity at airports, resulting in the airports increased inefficiency. An increase in the number of ground-handlers would result in a reduction in security at airports as, 'More door keys, access cards, [and] security passes would have to be issued, more personnel to screen...'[73] The 'problem' of technical feasibility focused on the practicality of allowing competitors into the market. For example at Frankfurt, it was stated, 'it would not be possible to allow the baggage system to be operated by a number of handlers without a significant loss of efficiency.'[74] Airports would also face increased costs from liberalisation in order to maintain safety and security, according to ACI.[75]

The German airports group ADV stated that it wanted to limit the amount of competition on the airside. A limitation of such competition would protect employment in a sector that is relied on heavily by the Germans; this resulted from the historical situation at German airports.[76] As stated above, ADV were supported by the German union ÖTV, and job protection was a major cause of concern both for ÖTV and ADV. However, the protection of jobs, although a legitimate concern was perceived as insufficient in itself to present to DGIV as a reason to limit competition. Therefore, it seems that the reasons of safety, security, cost and technical feasibility, in addition to the social / economic impact of liberalisation, were put forward by ACI as reasons to limit the extent of ground-handling liberalisation.

The February 1994 Document also detailed the limiting factors of liberalisation in regard to the alleged space constraints at airports. Lack of physical space was the problem which airports pointed to as the primary limiting factor preventing airside competition. ACI and ADV member Frankfurt however offered the firmest position on this issue:

> Capacity constraints make it impossible to increase the number of handlers. Only the exclusive provision of airside ground-handling ensures a high quality level of service.[77]

In terms of employment, the stated negative effects of ground-handling liberalisation were highlighted by ACI; it stated that liberalisation would result in an automatic loss of employment at the majority of airports surveyed.[78] The greatest amount of detail in this regard was reserved for Dusseldorf Airport where exacting German employment conditions were highlighted. ACI stated that German employees were subject to a number of pieces of legislation, including the 'Federal Collective Agreement for Public Employees', which broadly states that the contracts of employees over the age of 40 are not subject to termination if the employee has been employed for fifteen years or more. ACI stated: '...any release of ground-handling services would lead to substantial unemployment', despite the fact that other suppliers would recruit a 'small' number of employees. Most employees would be 'left without a job' and would become 'permanently unemployed'. Any redundancies would be subject to at least five pieces of legislation, including the 'German Dismissals Protection Law'. Airports would have to grant 'severance payment', however, 'social hardship' would result in every case, claimed ACI.[79]

The Vice Chairperson of Frankfurt Airport, Manfred Schölch, was a member of the *Comité des Sages*; a Committee established by the European Commission, 'to reflect on the future of aviation in Europe as an essential tool for economic and social development.'[80] However, the *Comité des Sages* was established to provide a level of credibility to the Commission, and to show that the Commission was 'listening' to the air transport industry. Its views would, therefore, carry considerable weight.[81] Its report, entitled 'Expanding Horizons' was written from June 1993, and published in January 1994, five months after the Commission Press Release on ground-handling (which suggested a *Commission* Directive) but in advance of the Commission Proposal for a Council Directive. Other members of the Committee included the former chairperson of an aerospace manufacturing company, a former minister for transport, and the chairpersons of Lufthansa and KLM Airlines (both AEA members). As well as the vice chairperson of Frankfurt Airport, the Committee also included the Chairperson of the French union *Force Ouvrière,* as an air transport union representative on the Committee.

The latter two members produced a 'Dissenting Opinion' from the majority of the Committee on the issue of ground-handling, under the section of the report, 'Ensuring the Right Balance - Airports and Air Carriers as Partners.' This issue was one of only 2 Dissenting Opinions in a total of 24 topic areas where recommendations were made in the report.

However, such was the relevance of this Dissenting Opinion (despite the fact that less than one fifth of the Committee supported it), that it is highlighted in the eventual Commission Proposal for a Council Directive.

The majority opinion of the Committee stated: '...ground-handling services at European airports must be fully liberalised as soon as possible.'

Adding:

> Airports may seek exclusion to or limitations of competition in ground-handling services only if they can demonstrate that, if they did not, the optimal use of their capacity would be hampered.

The Committee therefore stated that the *raison d'être* for the exclusion from ground-handling liberalisation should be extremely limited, and also clearly identifiable. The Committee added,

> ...airport operation must seek to ensure unbiased provision of services...and extend airport facilities in line with airport needs.[82]

Thus, the Committee clearly stated the necessity of preventing airports from using discriminatory tactics against ground-handling competitors (e.g. placing their competitors' operations a greater distance from terminal facilities), and the need to allow airports to expand to cope with changing demand, thus challenging the 'space constraint' argument used by the opponents of liberalisation.

By contrast, the Dissenting Opinion states that airports should be seen as 'public institutions providing infrastructure to aviation'.[83]

> The Commission may *urge* (emphasis added) European airports to liberalise ground-handling...only if local conditions allow this without severe negative effects on their functions as an infrastructural institution, or on their ability to invest according to the needs of further development of aviation and airport's competitiveness.
>
> Under all circumstances, airports must remain in a position to ensure safety and security of their operations by licensing all service providers.[84]

The Dissenting Opinion is extremely shrewd in two ways. Firstly, it clearly displays that the Committee was divided (albeit by a majority to a minority), yet the division is clearly recognised. Secondly, it displays a solidarity between the main German airport (the position which would have been formulated prior to the Committee report through ACI), and the main French union associated with air transport. Policy formulators were therefore aware that too drastic a liberalisation could potentially face substantial union resistance from both ÖTV as well as Force Ouvrière. It would appear that there is more than a hint of similarity between the Dissenting Opinion and the position of ACI. The Dissenting Opinion stated that the Commission may only urge ground-handling liberalisation. Indeed, a Commission Directive based on Article 90, so opposed by ACI, would stipulate minimum criteria that airports would need to adhere to. It would go beyond urging and establish a minimum statutory requirement for the liberalisation of ground-handling. Such a statutory requirement arising from a Commission Directive, based on an Article 90 Directive would violate the principle of airport control put forward by the Dissenting Opinion and by ACI.

Landside-handling

ACI agreed 'in principle' to the liberalisation of landside services,[85] subject to 'space constraints' within the airport;[86] i.e. ACI did not oppose this measure, but felt that the overriding factor is that the airport must be in control of the situation, and that the airport remains 'sovereign' over such issues.[87]

Summary of ACI position

A broad guide to the views of all the concerned interests can be obtained from the final Commission Proposal for a Council Directive on ground-handling:

> In general air carriers and ground-handlers favoured in varying degrees the guidelines put forward [in the Proposal]. Airports and trade unions expressed opposition to too full and too abrupt a liberalisation. Many replies stressed the need to bear in mind the social problems which reform of this sector could generate in the light of existing situations.[88]

ACI utilised a combination of tactics to lobby for this 'minimalist' position expressed by airports and supported by trade unions. It warned against the use of a Commission Article 90 Directive, which might be formulated primarily by DGIV. It saw such a Directive as 'improper' to say the very least. ACI also specified criteria for a Council Directive; essentially a 'minimalist' position on liberalisation that stemmed from the demands of ADV. Practical matters that would hinder liberalisation such as lack of space, security, safety, cost efficiency, technical feasibility in addition to the 'negative' effects on employment (and the alleged legal problems in Germany resulting from such unemployment), were emphasised. In short, ACI demanded no 'minimum number' of ground-handlers. It stated that the ownership of the ground-handler (such as the state), was irrelevant, and that airlines should have no automatic ability to handle on the airside.

To some commentators such reasons are an 'excuse' to avoid ground-handling liberalisation.[89] As argued in Chapter 4 it was ADV which arguably played the greatest role both in the formulation of the position of ACI (Level II interest representation, via the interest group), as well as in its direct contact with the European Commission and European Parliament (Level III). ADV also played the greatest role in targeting and utilising its Member State to achieve its policy goals (Level 1). At Level II ADV played an extremely prominent role via the ACI European policy committee. Crucially, resulting from the structure of ADV's decision making, ADV always had the support of its members Frankfurt and Dusseldorf, themselves 'independent' members of the ACI European policy committee.

The positions of the Independent Interests

Virgin Atlantic was not involved in attempting to influence the outcome of the ground-handling Proposal. This was for two main reasons.

The positions of Director of government and external affairs and manager of aeropolitical affairs were not created until 1995 and 1997, respectively. Thus, until 1995, Virgin's influence was very limited at all three levels of interest representation; at the state, via an interest group (as Virgin is not a member of AEA or ERA), and directly to the Commission and European Parliament. As in the case of the majority of regional airlines, it was the chairperson that was relied on for the airline's lobbying activity.

Secondly, all of Virgin Atlantic's EU activities took place from Heathrow, Gatwick and Manchester. All three airports enjoyed, and continue to enjoy a high level of ground-handling liberalisation. As Virgin Atlantic did not operate to destinations within the European Union[90] (other than the UK), it might be argued that the issue was of little concern to Virgin. However, Virgin Atlantic stated after the formulation of the final Directive that they wished there had been a stronger Directive which further liberalised the EU ground-handling market.[91]

Servisair, as an independent ground-handler, is an obvious supporter of ground-handling liberalisation. Any liberalisation of the market would result in more opportunities for Servisair to conduct their activities. Subsequently, Servisair called for the complete liberalisation of ground-handling on the land and the airside. Servisair fully supported the concept of a choice of third-party handlers and wanted the choice to be as great as possible in order to allow them to enter the market at various EU airports. Like ERA, Servisair opposed any minimum criteria that airports could utilise to 'avoid' ground-handling liberalisation.[92] The commercial activities of Servisair were (and are), being limited outside the UK because of the restricted nature of the ground-handling market.[93]

However, as detailed in the previous chapter, although Servisair is the largest firm in its sector, it is extremely limited when compared in terms of overall size to other air transport interests. Moreover, although the views of Servisair were shared by the only other comparable firm in the European Union (Belgavia)[94] its activity at all three levels of interest representation was, and is, extremely restricted. Despite support from the UK government,[95] Servisair had little success in attempting to persuade the Commission or European Parliament of its views.[96]

Like the majority of non-elite interests, Servisair relied on its Chief Executive for its interest representation activities; it did not and does not have a particular department dealing with such matters.[97] Therefore, lack of executive time was and is a considerable problem and its representation at all three levels was and is extremely restricted. Moreover, as an Affiliate Member of ACI (see above), Servisair could not participate in the

European policy committee of ACI unless invited to do so, which, according to my knowledge, it was not.

Slot-allocation

As stated in Chapter 3, 'Slot congestion' occurs where the user of a slot at a 'preferred time' or 'preferred airport' has a commercial advantage over another airline that seeks access to such a slot, but is unable to gain access. There is no exact numerical level where an airport becomes 'congested'. At Manchester Airport (annual traffic units: 15,317,438),[98] the airport is considered congested at 'peak periods' in the morning and evening. At other times of the day the airport is relatively free of congestion.[99] The Chief Executive of Luton Airport in the UK (which has a relatively small traffic figure of less than two million units per annum),[100] stated in January 1997 that he did not think that the airport would face 'congestion problems' until the airport operated at over twelve million passengers per annum.[101] However, by August of that year certain airlines which operate from Luton were voicing concern that the airport was becoming congested at various times of the day[102] (this shows the speed at which successful airports can increase their popularity). Conversely, the report of the European Parliament's Economic and Social Committee of 1991 highlighted only six airports as 'congested' where there was a slot-allocation problem - London (Heathrow, and Gatwick airports), Munich, Dusseldorf, Frankfurt and Milan (Liante).[103] Since that time it has become clear that slot congestion is a potential problem at airports which traditionally have not been considered as congested. This is because airlines have become increasingly concerned as to the timing of a slot, rather than simply its availability throughout the day. In short, an issue which mattered little to regional airlines and non-elite interests at the beginning of the 1990s has become an issue of increased concern to airlines across Europe, and not only to those airlines which have sought access to traditionally congested airports. Furthermore, as airlines grow in size, slot congestion becomes an issue which begins to affect airlines that it did not previously (such as Gill Airways, noted in Chapter 4).

The issues of contention in the revision of the 1993 slot-allocation Regulation Proposal

Two major issues of contention surround the formulation of the Proposal for the revised Regulation: 'timed historics' (or 'grandfather rights'), and secondary slot trading.

Grandfather rights allow an airline to continue to take-off or land at a particular time indefinitely, subject to certain safeguards. Grandfather rights are recognised in the first slot-allocation Regulation (95/93), in that compulsory surrender of such rights only happens if the slot is not operated on at least 80% of the occasions in any given season (commonly known as the 'use it or lose it' rule).[104] However, the airline may operate to any destination that it sees fit at this particular time. Grandfather rights are not unique to the EU and are practised in most states world-wide. Those who oppose grandfather rights might argue that the world-wide recognition of such rights is reflective of the anti-competitive nature of international air transport. Grandfather rights also feature as part of the IATA guidelines when slots are scheduled at the bi-annual IATA conferences, which all scheduled airlines attend. Those holding grandfather rights may continue to operate at the particular times from the particular airports at which they operate, unless they wish to surrender the rights.[105] The co-ordination of slot timings began in the 1940s when airlines wished to ensure 'connections', particularly on transatlantic routes, which may have required a change of aircraft.[106] However, it is imperative to note that states (or a group of states such as the EU), may dispense with the grandfather guidelines if they so wish. The precedent for this was the withdrawal of grandfather rights by the US Federal Aviation Administration (FAA), from airlines at certain airports. The stated aim of the FAA was to enhance competition at these airports.[107]

The ability of an airline to trade a slot on a secondary market is, essentially, the ability for an airline to trade, at a mutually agreeable price, its grandfather right. For example, if Air France were to 'purchase' the right from British Airways to takeoff from Heathrow at 0900 daily then that right would pass to Air France indefinitely (subject to the 'use it or lose it' rule). The seller of the right may set any price that it sees fit, and it is for the buyer to meet its price. Regulation 95/93 makes no mention of the legality or otherwise of airlines to engage in a secondary slot market. Legal opinion is divided as to whether Regulation 95/93 allows the activity to happen (see Chapter 6). The view of British Airways, the most vocal supporter of the ability of airlines to engage in this activity, is that as Regulation 95/93 does not prohibit the activity it is therefore lawful.[108] However, according to the views of both DGIV (Competition), and DGVII (Transport), secondary slot trading is illegal under Regulation 95/93,[109] although DGIV and DGVII fundamentally disagree as to whether the revised Regulation should clarify the status of secondary slot trading as a legal *bona fide* activity.[110] In summary, the situation is one where interests either support, or are against, slot trading and / or grandfather rights.

The position of the Association of European Airlines

Grandfather rights

The airlines of the AEA fully and openly support the continuation of grandfather rights, and have publicly communicated this view to the Commission.[111] This position stems from the fact that the flag carriers were (and normally are), the main airlines based in the main airports of their country of origin. For example, British Airways possess the single greatest number of grandfather rights at Heathrow, Air France possess the single greatest number of grandfather rights at the main Paris airport, Charles de Gaulle, Olympic Airways dominates Athens, etc. Therefore, any compulsory surrender of these rights would, in the view of these airlines, be detrimental to their interests. The official view of the AEA is that the continued access by its airlines to their home states' main airports (or hubs as they are described), is crucial for air transport 'networks' to develop in Europe. Any confiscation of grandfather rights would therefore reduce the flag carriers' ability to operate such networks.[112]

Secondary slot trading

Officially, the AEA, 'has never discussed [secondary] slot trading.'[113] The only AEA member to have publicly declared its total support for secondary trading is British Airways, arguing that it is the only way in which congested slots can be freed at European airports.[114] However, the evidence suggests that the airline members of the AEA may well have engaged in Level III lobbying (directly to Brussels), in order to promote secondary trading. It is likely that the AEA has also unofficially and privately stated to DGVII of the Commission that it supports secondary trading but the airlines of the AEA cannot admit publicly to their support of the concept - to do so might be seen as an admission of the fact they already engage in the practice. A senior source within DGIV stated that DGIV is investigating airlines that engage in the practice, which it considers illegal.[115]

At the outset it must be noted that the AEA airlines do not oppose secondary slot trading. A commentator has noted that such non-opposition to this concept (particularly a concept which has so openly been discussed before a Proposal has been formulated), is tacit support. The clearest indication of this is a senior source within DGIV who confirmed that it is the flag carrier airlines which support the clear legalisation of secondary slot trading in the revision of the Regulation (DGIV opposes any such legalisation as anti competitive).[116] As the AEA members possess the greatest number of slots at Europe's main (and therefore most congested) airports, secondary trading would be to their enormous advantage, as the slots held would become official fixed assets of the airline. The value of a pair of 'peak' slots is not clear (i.e. one take-off and one landing slot at a hub airport at times convenient to the passenger), but it is estimated that pairs of slots have been traded for US$10 million on the current 'grey market' (a market of which the legal validity is unclear), between airlines.[117] This figure is purely an estimation, as slots have never been openly and accountably traded, although the evidence is overwhelming that such trades have, and continue to, take place, colloquially known as 'under the table' trading. It is clear that the flag carrier airlines, who dominate the hubs would have assets of substantial value if the practice was to become clearly lawful.

DGVII commissioned consultants Coopers and Lybrand to suggest possible modifications to Regulation 95/93.[118] According to the author of this Coopers and Lybrand report, the work was carried out in a 'policy vacuum' as the Commission produced no set of policy objectives to assist the execution of the study.[119] Nonetheless, the report found in favour of the legalisation of secondary slot trading.[120] In this report all European Union AEA member airlines (with the exception of Aer Lingus, British Midland

and Luxair)[121] were consulted.[122] No ERA member was consulted despite the fact that 70%[123] of ERA members' routes are to or from a hub airport. The report subsequently favoured secondary trading,[124] evidence perhaps of a link between the input of the consulted airlines and the outcome of the report. Furthermore, in October 1996 a 'draft' Commission Proposal was leaked by a source within DGVII to air transport interests. It stated:

> After careful consideration of all other options to stimulate the necessary flexibility at highly congested airports, the Commission has opted to facilitate exchanges of slots between air carriers by removing the obligation to give a slot in return for a slot, and by allowing the monetarised trading of slots between air carriers.[125]

British Airways vocally supported this draft Proposal, and no AEA member airline opposed it. Indeed, the opposition to this Proposal came primarily from DGIV (as examined in Chapter 6).[126] The following November a source within DGVII publicly stated that the findings of the Coopers and Lybrand report would: '...form the basis of the forthcoming Proposal.'[127] Again, no opposition to this fact was put forward by any AEA member.

The support by the AEA of grandfather rights is clearly stated above. The support by the AEA of secondary trading is implicit in its non opposition. At any time during the course of the attempted formulation of the Proposal the AEA could have declared its opposition; to the knowledge of the author, it did not and has not done so. AEA members already partake in secondary trading, which the Commission (both DGIV and DGVII), considers to be illegal. The evidence for this trade is clear, but open support by the AEA for secondary trading might well be seen as an admission that it 'officially' knows secondary trading occurred. Only British Airways has publicly declared its support for slot trading. The reason why it is only BA that has stated its support may be as a result of the conditions imposed by the UK government (and the much harsher conditions by the European Commission), if British Airways was to pursue an alliance with American Airlines. The UK government initially demanded that British Airways 'surrender' 168 grandfather rights per week at Heathrow Airport. British Airways insisted that if such a forced surrender took place that it must be able to sell the slots on the open market.[128] British Airways were subsequently 'forced' into a position where it had to openly support secondary trading. It might be argued that the other flag carriers of the EU could afford to 'free ride' the public view of British Airways' position.

Overt support for secondary trading is complicated by the fact that, despite the fact that the Commission has stated that secondary trading is illegal, many airlines, in particular the AEA airlines (with greater financial resources), have 'covertly' taken part in secondary slot trading *prior to* any explicit legalisation legalising the practise.[129] This situation of 'grey market' or 'under the table' trading significantly complicates the stand on slot trading taken by the groups examined. The majority of airlines (the notable exception being British Airways), feel that to overtly support the concept of slot trading would be a tacit admission that they themselves already engage in the practice.[130]

The position of the European Regional Airlines Association

ERA officially support the concept of grandfather rights and has stated this clearly to DGVII. ERA state that to remove such rights would discourage investment by airlines in particular routes, as the airlines would have no guarantee that their investment in the route could continue.[131] This position has been maintained throughout the attempted formulation of the Proposal. A source from one of the smaller members of ERA has however stated that several airlines, in principle, oppose grandfather rights, because they do not have the access to the hub airports that some of the larger airlines have.[132] This position is not therefore a unanimous one throughout the membership of ERA. That ERA supports grandfather rights is not and indeed cannot be due to the dominance of ERA members at hub airports (as it arguably is with AEA airlines), but due rather to the fact that if a regional carrier has managed to establish even very few slots at a hub airport, it will be keen to continue to possess these slots. Despite the fact that 70% of regional services operate into hub airports[133] (note, this is entirely different from a 70% dominance of regional airports), the percentage of slots that a regional airline possess at a single hub airport is tiny in comparison to the flag carrier.[134] However, both prior to and post the streamlining of decision making within ERA, the principle of supporting the indefinite holding of grandfather rights has been adhered to.

Conversely, ERA's position on the secondary trading of slots has altered substantially. This change seems to have come about at the same time as the streamlining of decision making within ERA. In April 1996, in a presentation to Neil Kinnock the EU Commissioner for Transport, ERA stated:

> Regional airlines oppose any proposals which would result in the introduction of monetarised [secondary] trading. [It] would favour airlines with the deepest pockets to the detriment of Europe's peripheral communities.[135]

After the streamlining of decision making and the appointment of an air transport manager (highlighted in Chapter 4), the air transport manager stated:

> Some of our members have said that secondary trading does not adversely affect airlines which do not wish to sell their slots. Whilst we have not formally discussed it and come to a new position we are no longer saying that we are opposed to slot trading...we no longer oppose it in Commission meetings.[136]

There has, therefore, been a substantial switch in the position of ERA on this issue. It may be argued that this happened because of the greater attention paid to the needs of the subsidiary airlines within ERA. The position has therefore changed from a position of strong opposition, to one where ERA now states:

> Some of our members say it does not matter to you, if there is no obligation to trade, if other airlines are selling their slots. Therefore, if secondary trading is allowed it is a situation that is neutral at worst....[137]

The position of Airports Council International

When questioned, airports, on the whole, see themselves as the legal owners of slots,[138] as the individual airport has created the slot for airlines to use. However, rather confusingly, ACI does not oppose the ability for airlines to trade these slots.[139]

One legal view is that for airlines to claim 'ownership' of the slots is an 'economic anomaly' as the *airport* is the creator of the slot and therefore the legal owner.[140] An alternative view is that the legality of ownership is 'irrelevant'. Of relevance is whether it is the airline or the airport which has the ability to trade the slot.[141] ACI do not directly support the ability for airlines to secondary trade slots. 'We had a meeting in March 1997 to try and establish a new position on slot trading, but it proved impossible'.[142] ACI has, however, established five official positions on the matter of slot-allocation (rather than trading).[143] On the matter of grandfather rights, ACI can be described as 'broadly supportive'. ACI is against the 'confiscation of historical rights'[144] (i.e. the forced surrender of grandfather rights by the airlines). However ACI has recognised previously that grandfather rights, '...on the one hand provide the necessary stability in the planning process, on the other may impede market access to new entrants.'[145] Individual airports questioned on this matter vary in their support for grandfather rights. As stated in Chapter 4, ADV sees airports as pieces of transport infrastructure, vital to the economic well being of the particular region that the airport is based in. ADV has stated: 'in principle we support grandfather rights'.[146] BAA 'fully support grandfather rights'[147] whilst Luton Airport (an example of a relatively small airport), has stated:

> We see both sides of the argument. Grandfather rights can be seen as a barrier to competition. However, if an airline has invested heavily it would be unfair for a body such as the government to take them away. I can see both arguments.[148]

The view of a sample of airports on the matter of secondary trading is equally unclear. Of those questioned only BAA fully supported the ability for airlines to engage in secondary trading.[149] However, no airport questioned opposed the concept. ADV stated that the situation was, 'difficult, ...this situation takes place already, illegally'.[150] Manchester Airport highlighted the cost to themselves in the investment required to build its second runway, due for completion in May 2000. It stated that the airport should have, at least 'some interest' in the trading of slots, but admits that the majority of flag carriers support the concept of being able to secondary trade the slots themselves.[151]

It can be argued from a 'philosophical' stand point that ACI should oppose grandfather rights and secondary trading. The airports have built the infrastructure which allows the airline to operate the particular slot. Any trading of these slots should therefore financially benefit the airport (at least in part), as well as the airline. This is certainly the view privately shared by many of the smaller airports. However, the dominant view seems to be a reluctant acceptance that it is the airlines which bring in the revenue to the airports. Airports which are of both the 'liberal' camp and the 'non-liberal' camp on the issue of ground-handling appear to take this single view on the issue of slot-allocation.

This position of the airports on slot-allocation may seem surprising. If slots are potentially worth a considerable amount when traded, an observer might question why it is the case that the airports do not make a more vehement claim to the ownership of the slots. An answer to this question may be derived from the evidence presented by Matthew Hudson, Chairman of Prestwick International Airport, to a UK House of Commons Transport Select Committee enquiry.[152] In his response to a related question he attempted to separate UK airports into relatively large (such as BAA), and relatively small (such as Luton). Large airports, he stated, are able to derive enormous revenues from the fact that flag carrier airlines, in particular the home flag carrier (such as British Airways), operate from its airports. Indeed, Mr Hudson is correct. Although the majority of BAA's revenue derives from retail activities, it would be unable to get such revenue if the flag carriers (which use large aircraft and hence throughput many potential customers), withdrew their services in favour of an alternative airport. Therefore, it is in the long term interests of airport operators such as BAA not to 'upset' flag carrier airlines, particularly the home flag carrier. Small airports, such as Luton, are in a similar position, but for a different reason. An airport such as Luton wishes to attract as many airlines as it can, not only to increase the profitability of its operation, but also because the arrival of new services may increase the prosperity of the surrounding area.[153] Relatively small airports, on the whole, would not publicly admit that they are against grandfather rights or against the clear legalisation of secondary trading, as such statements may prevent a new carrier from entering the airport. Indeed, such comments might persuade an existing carrier to discontinue operating from the airport, as it might suspect that their grandfather rights could be withdrawn by the airport's owners.

Airlines (as well as passengers) are the customers of airports, and airports competitively battle with one another in order to attract airlines to their particular locations.[154] Hence, ACI, has subsequently kept a 'low profile' on this issue. From an academic / theoretical point of view it might well be argued that airports should indeed oppose grandfather rights and the ability of airlines to secondary trade. The reality of the situation, however, dictates that, in the main, airports are relatively neutral on these particular issues.

The position of Independent Interests

Virgin Atlantic has been, arguably, the most vociferous opponent of grandfather rights. However, its group, ACE, seems to have taken no position on the matter.[155]

Virgin's opposition to grandfather rights primarily stems from the fact that, in its view, the flag carrier airlines dominate the hub airports, to which access for 'new entrants' is vital, if successful competition is to take place. At its main airport of operation, Heathrow, Virgin possesses only 1.9% of the slots but seeks substantially more. Virgin claims that the grandfather rights system prevents them from gaining more slots.[156] Virgin states that slot availability is 'crucially important' to competition. Without access to slots airlines are unable to challenge the incumbent, dominant carriers.

> The dominant airlines, such as British Airways at Heathrow, and increasingly at Gatwick, and Lufthansa at Frankfurt, possess large slot holdings...because they inherited them from the days when they were their governments' favoured airlines.'
>
> Any proposal to amend the grandfather rights principle will therefore attract strong opposition from those airlines which have benefited from it, and no doubt their governments will oppose amendments also.[157]

Virgin proposes a 'thorough and objective review' of grandfather rights; airlines should have possession of a slot time limited to ten years. Once the right has expired the slot is allocated, not on the basis of an historic right, but by a system that 'reflects the quality and competitiveness of services offered by individual airlines'[158] (although Virgin is far from clear as to how such 'quality and competitiveness of service' should be evaluated).

Virgin argue that the reason why governments accept the concept of grandfather rights (highlighting the case of British Airways), is that the flag carriers are highly effective at lobbying the institutions of government, civil servants and back benchers. This has resulted in a situation where the distribution of slots is 'completely meaningless' and where the grandfather rights held by the flag carriers equate to the type of direct financial subsidy that Air France *et al* are so criticised for receiving.[159]

When the AEA, ERA and ACI speak of secondary trading, they refer to secondary trading as the ability to trade the grandfather right. Virgin is opposed to grandfather rights and therefore opposed to any trade of the grandfather rights.

> It has been suggested by some that the problem of slot scarcity can be easily solved by allowing slots to be traded openly. This is not true. There may be benefits in permitting open slot trading, although there are also dangers as US experience has shown, but it is naive to argue that the problems experienced at airports such as Heathrow can be solved in this way. Informal slot trading has taken place for many years, with no real effect.[160]

Virgin argue that if slot trading is to take place, revenue from the slot should go to the exchequer, or to the airport (as the creator of the slot), rather than the airline; this is the opposite view to that of the flag carriers. Additionally, Virgin argue there must be a cap on the amount that can be charged for a particular slot.[161]

Followers of air transport news may find this position a little odd. It is widely rumoured that Virgin wish to bid for the regional airline 'Cityflyer Express' a Gatwick based regional airline, which operates as a franchise of BA. Some argue that Virgin would 'asset strip' Cityflyer if it won control, and use Cityflyer's prime slots for Virgin's lucrative routes.

Virgin may therefore have no use for the 650 staff currently employed by Cityflyer.

Servisair plays, by its own admission, no role in the attempted revision of the EU slot-allocation Regulation.[162] However, Servisair would clearly benefit from increased access by new entrant airlines to Europe's hub airports, as this increased access would provide potential custom for Servisair, as new entrants would not be compelled to use in-house or 'preferred' ground-handers in the way that many of Europe's flag carriers currently do. The way in which this access should be increased, either by the retention of grandfather rights and the ability of airlines to secondary trade, or by the abolition of grandfather rights put forward by Virgin, it not a view on which Servisair currently take a position.

Summary

The elite group of flag carriers, the AEA, presented the most diluted (weak) position on ground-handling. A small number of flag carriers were vociferous supporters of liberalisation; an equally small number were firmly against liberalisation. Some carriers, as a result of their 'cross ownership' (i.e. their government owns both the airline and the airport), took positions against their own governments. The AEA, although consisting of elite interests, was unable to formulate a firm position because of the dissension within itself, and its inability to produce a decision based on majority (rather than unanimous), voting. The AEA suffered from an 'ostrich mentality' in the formulation of the ground-handling Proposal. It was the last interest group to produce an official opinion which, when produced, was the broadest and the weakest of the three groups examined. ERA were the firmest supporters of ground-handling liberalisation, supporting full self-handling rights and the ability of any firm to enter the ground-handling market and supply a competitive choice for the airlines. ERA also stated that it wished to see no minimum size of airport set for the application of the liberalisation. Yet ERA was pragmatic in that it stated that if there was to be a minimum it should be no more than 1 million passengers *per annum* (although, some might argue that ERA weakened its bargaining position by stating this).[163]

In the case of slot-allocation, the AEA openly support the retention of grandfather rights, whilst it does not oppose the clear legalisation of secondary trading. Although the airlines of the AEA appear to support this, they appear to be inhibited in openly declaring their support because of the 'illegal' situation of secondary trading which already takes place. ERA support grandfather rights, but have changed positions on the matter of secondary trading. ACI support the retention of grandfather rights, but has been unable to formulate a position on the matter of secondary trading of slots. Clearly, many airports could not individually decide whether or not they favoured the concept. Virgin Atlantic were the clearest opponents of grandfather rights, and were against the secondary trading of grandfather rights. Although, it must be noted that as Virgin acquire more slots of its own at Heathrow and Gatwick that it is rapidly starting to change its tune.

Notes

[1] See: Corinne Dussart-Lefret and Christine Federlin, 'Ground-handling Services and EC Competition Rules', in *Air and Space Law* (Vol. XIX), No.2, 1994, pp.50-61; Wolfgang Deselaers, 'Liberalisation of Ground-handling Services at Community Airports', in *Air and Space Law* (Vol. XXI), No.6, 1996; pp.260-266; Ian Stockman, 'A Comparative Study of User Costs at Selected European Airports' (Air Transport Research Group), Cranfield University, 1996 (for comment, see: Ian Stockman, 'Ramping up the price', in *Airline Business,* Vol.12 (Nov, 1996), pp.53-58).
See also: UK, Dept. of Transport, Aviation Group, EC Aviation Liberalisation Information, Summer 1996, p.15.

[2] Several commentators differ in their terminology used to describe this separation. A more accurate, although less used description are the terms, 'passenger handling' and 'ramp handling'. The US term for the activities are: 'under wing' and 'over wing' (interview: Anthony Rowbotham (General Manager, Business Development), Servisair, Manchester, 16.7.97).

[3] European Commission: 'Proposal for a Council Directive on access to the ground-handling market at Community airports', COM(94) 590*final* (Annex), p.13.

[4] COM (94) 590*final* (Annex), pp.13-14.

[5] Interview: Servisair, Manchester, op.cit.

[6] Airlines prefer to be able to choose who operates their ticket desks and handles passenger queries, as these personnel represent the airline to the passenger. Airlines do not want to risk their image on check-in staff (who may be slow, rude, or downright abusive) whom they have no control over.

[7] Le Thi Mai (Head of Infrastructure and Environment, AEA), 'Ground-handling liberalisation', paper presented at the IBC, 'European Air Transport Services and Airports', conference, Brussels, 6-7.2.97. However, a survey of ERA airlines, a proportion of which operate from airports outside the EU, revealed that 54% of airports provide a

monopoly on the landside in addition to the airside (source: ERA, 'Ground-handling in Europe', 22.2.95).

[8] Only 4 airports in the EU offer a choice of at least three ground handlers on the airside, (that is a choice which does not exclusively include the airport and / or the flag carrier). These airports are: Manchester, Heathrow, Gatwick (UK), and Schiphol (Amsterdam, Netherlands), Source: AEA, 'Yearbook', 1995, p.23.

[9] Interview: Steven Aylward (Manager, Competition and Regulatory Affairs), British Airways, Heathrow, 6.9.96.

[10] Interview: Tim Walden (Manager, Commercial and Industry), British Midland Airways, London, 16.1.97. However, this British Midland representative was referring to an official position of the AEA. Like the ERA, examined below, the AEA did not produce an official position until after the Commission had formulated its Proposal.

[11] DGIV Press Release, IP/93/714, 25.8.93.

[12] Although, at several Spanish airports Servisair has established limited 'supervisory' services. Servisair has been unable to establish any such services in Greece (see Chapter 4 for a description of supervisory ground-handling services).

[13] Interview: Servisair, Manchester, op.cit.

[14] See: ERA, 'Ground-handling in Europe', 22.2.95, p.6.

[15] ACI Europe, 'Consultations on Ground-handling Services. Response by ACI Europe to the European Commission DGIV and DGVII' (Feb, 1994), p.6-7.

[16] Christian Blanc, Chairman of Air France was dismissed by the French Transport Ministry following pressure, particularly from the French public service unions. Blanc was attempting to 'streamline' the finances of Air France beyond what the public service unions were willing to accept. The example of Christian Blanc is not unique. The Chairman of Olympic, Professor Regis Doganis may have been dismissed by the Greek government for attempting to modernise Olympic and cut the airline's expenses in a way that was not acceptable to the unions or the then Greek government. Such control of state enterprises fully supports the arguments of Chapter 4, detailing the 'intimate' relationship between the flag carrier airline and its government, particularly when the flag carrier is majority state owned.

[17] Often it is the largest airport within a state which is cited by critics as the airport that supplies a monopolistic service. For example, all the airports in Germany provide a monopolistic service (on the airside). However, critics will normally point to Frankfurt, the largest German airport, as a prime example. Similarly, Madrid, Milan, Rome and Athens are pointed to as monopolistic ground handlers, whilst their behaviour is mirrored by the other airports within these states.

[18] In the case of SAS this was the support of Denmark, Sweden (yet to be an EU member at that time), and Norway (which subsequently rejected EU membership).

[19] As in the majority of AEA decision making, the matter was of the greatest relevance to EU members. However, airlines of non EU members such as Cyprus Airways fly into airports (such as Frankfurt) where airport supplied ground-handling monopolies exist. Such airlines were in the advantageous position that they would benefit from a

liberalisation of the market (if a liberalisation was to occur), without having to undo their own ground-handling monopolies, as they are outside the remit of the EU / EEA. Austrian Airlines, at the time of the formulation of the Commission Proposal was not an airline based in the EU (although it was in the EEA). However, like Alitalia and Air France the owner of the airport, and the airline was the state. Austrian Airlines was not a signatory to the original complaint to DGIV.

[20] Policy making can, arguably, be divided into two main areas. First, there exist the technicalities of the policy position. For example an airline might request a more liberal ground-handling market, as this it claims, would lead to lower air fares and greater competition. Second, there are the 'political dimensions' of a policy. For example, the political pressure that can be applied by a particular interest group. The two aspects of policy making although related, can be described as 'semi-detached'. It is revealing that the 'experts in the field' of the AEA seem to have been aware of such 'political dimensions' when they were promoting their views. This situation is in stark contrast to that which exists within ACI. Policy positions seem to be formulated within ACI by political experts rather than 'technicians'. (Interview: Sheena Gibson (Head of European Policy Unit), ACI, Brussels, 10.4.97.)

[21] Interview: Kees Veenstra (Deputy Secretary General), AEA, Brussels, 16.4.97.

[22] Ibid.

[23] See: Andrew M.McLaughlin, Grant Jordan, William A.Maloney, 'Corporate Lobbying in the European Community', in *Journal of Common Market Studies,* Vol.31, No.2 (Jun, 1993), pp.191-211.

[24] Interview: British Airways, Heathrow, op.cit.

[25] A Directive based on Article 90 of the EC Treaty allows the Commission to form a Directive *without* having to achieve the support of the Council of Ministers. See Chapter 2 for further details.

[26] Le Thi Mai (Head of Infrastructure and Environment), 'Ground-handling liberalisation', Brussels, op.cit.

[27] AEA, 'Yearbook', 1995, p.23.

[28] John Balfour (Solicitor, Frere Cholmeley Bischoff), 'The EC Ground-handling Directive Explained', paper presented at the Royal Aeronautical Society, 'Slots and Access to Airport Facilities' conference', London, 21.1.97. For references on franchising, see Chapter 4.

[29] Interview: British Airways, Heathrow, op.cit.

[30] Interview: Brian Simpson MEP (Member, Transport and Tourism Committee), Brussels, 20.4.98.

[31] COM (94) 590 final, 13.12.94.

[32] ERA, 'Ground-handling in Europe', 22.2.95, p.6.

[33] COM (94) 590 final, 13.12.94, p.19.

34 Articles 6 and 7, COM (94) 590 final, 13.12.94. See Chapter 6.

35 ERA, 'Ground-handling in Europe', 22.2.95, p.9-10.

36 For example, the airports that comprise *Airports de Paris*, give, seemingly, a choice of three ground-handling agencies; however, all are government owned. ERA stated that it was concerned that true competition would be stifled at such airports, as, for example, the French government could ensure that the companies did not compete on price.

37 An example highlighted is Brussels (Zavantem) Airport. The aircraft of Sabena (the flag carrier), are parked closer to passport control (hence a shorter walk for the passenger), than the other airlines that the airport-appointed ground handler (Belgavia), handles. This example is highlighted by: Trevor Soames (Partner, Norton Rose), in 'Access to Airport Facilities, The impact of Competition Law', paper presented at the Royal Aeronautical Society, 'Slots and Access to Airport Facilities', conference, London, 21.1.97.

38 This number of passengers included transfer passengers.

39 ERA, 'Ground-handling...', op.cit, p.12.

40 A Committee to oversee the operation and selection of ground handlers.

41 Article 5, COM (94) 590 final, 13.12.94.

42 ERA, 'Ground-handling...', op.cit, p.9.

43 For example, ground-handling represents 38% of total turnover at Frankfurt. (Source: Section 6, p.C.12, ACI February 1994 document.) Airports generate revenue in a number of different ways. All airports charge airlines a landing, takeoff and parking fee. As well as this many airports (for example, the airports of ADV), generate a substantial proportion of their revenue from ground-handling activities. Conversely, the airports of the BAA group in the UK (who have the most liberal ground-handling situation in the EU) choose to concentrate their revenue generation from retail activities. See: Barry Gibson (Group Retail Director, BAA), 'Airport Concepts for the Next Century - The Role of the Airport in the Total Consumer Experience', paper presented at the 'FT World Aerospace and Air Transport Conference', London, 29-30.8.96.

44 Interview: AEA, Brussels, op.cit.

45 Indeed, a less liberal ground-handling market at a German airport could only be of an advantage to BAA, as the financial cost to airlines to use BAA airports would be lower than the non-liberally minded German airports, such as Frankfurt.

46 ACI Europe, 'ACI Position on Ground-handling at Airports' (Executive Summary, ACI Europe, Mar. 1991).

47 ACI Europe, 'ACI Europe Resolution on Handling' (adopted by the General Assembly of ACI Europe), Madrid, 13.11.92.

48 DGIV: Press Release, IP/93/714, 25.8.93.

49 ACI Europe, 'ACI Europe Resolution on Ground-handling' (adopted by the General Assembly of ACI Europe), Malta, 10.6.93.

50 ACI Europe, 'Consultations on Ground-handling Services. Response by ACI Europe to the European Commission DGIV and DGVII' (Feb. 1994).

[51] ACI Europe, 'ACI Europe Resolution on Ground-handling' (adopted by the General Assembly of ACI Europe), Turin, 9.6.94.

[52] That being: ACI Europe, 'Comments of ACI Europe on the Draft Council Directive on Access to the Ground-handling Market at Community Airports' (May, 1995). Clearly, ACI recognises that the most effective lobbying is done at the pre-Proposal stage.

[53] Interview: Sheena Gibson (Head of European Policy Unit), ACI, Brussels, 10.4.97.

[54] This supports the view that an interest is most successful when it makes itself an 'indispensable source of information. See: R.Hull, 'Lobbying Brussels, A View From Within', in S.Mazey and J.Richardson (eds) *Lobbying in the European Community*, (Oxford: Oxford University Press, 1993), p.83.

[55] Interview: Sheena Gibson (Head of European Policy Unit), ACI, Brussels, 10.4.97.
See Chapter 2, for details of a Directive based on Article 90.

[56] DGIV Press release, IP/93/714, 25.8.93.

[57] Trevor Soames (Partner, Norton Rose), 'Ground-handling Liberalisation', paper presented at the IBC, 'European Air Transport Services and Airports' conference, 6-7 2.97.

[58] ACI Europe, 'ACI Europe Resolution on Ground-handling' (adopted by the General Assembly of ACI Europe, Malta, 10.6.93), p.1.

[59] ACI Europe, 'ACI Position on Ground-handling at Airports' (Executive Summary, ACI Europe, Mar. 1991), p.2.

[60] ACI Europe, 'ACI Europe Resolution on Handling' (adopted by the General Assembly of ACI Europe, Madrid), 13.11.92, p.1.

[61] ACI Europe, 'ACI Europe Resolution on Ground-handling' (adopted by the General Assembly of ACI Europe, Malta), 10.6.93, p.1.

[62] An Article 90 Directive.

[63] That is without the political control of the Council of Ministers.

[64] Point 13.2., pp.18-19 ACI (Feb. 94 Document), op.cit. This view was expressed as a point of law in a solicitors document, compiled by Scott Crosby (Solicitor), Partner, Kemmler, Rapp, Böhlke and Crosby, Brussels, 28.2.94.

[65] Point 14.4, p20, in ACI (Feb. 94 Document), op.cit.

[66] Refer to Chapter 1 for citation by Greenwood, on the 'mutual benefit' of developing 'clientilistic' relationships with the relevant parts of the Commission.

[67] Point 5.2, p.7, in ACI (Feb. 94 Document), op.cit.

[68] Point 7.2, p.9, in ACI (Feb. 94 Document), op.cit.
Several land-mark cases are cited to prove this point, claim the solicitors of ACI, including: *BEA v. City of Manchester* (judgement of the UK Court of Appeal, 3.10.66), and various cases in Italy, Germany and the Netherlands. See f.n. No.9, p.22, in ibid.

[69] Point 7.3, pp.9-10, in ibid.

[70] The legal basis for this claim is based on the opinion of the solicitor for ACI in the above document. The claim is based on the rejection by the ECJ of Article 8 of the European Commission Directive, on competition in the markets for telecommunications services (O.J. 1990 L.192/10) which was based on Article 90 of the Treaty. Article 8, which stated that existing binding contracts should be terminated, was rejected by the ECJ. A similar provision in the European Commission Directive, on competition in the markets in telecommunications equipment (O.J. 1988 L.131/73) based on Article 90 (3) was also rejected. Point 7.6, p.10, in ibid. See cases: C-271/90, *Services Directive,* 1992. and C-202/88, *Terminals Directive.*

[71] Point, 9.4, p.13, in ibid.

[72] See Section 1, and p.C2, in Ibid, for list of airports.

[73] Section 5, p.C.9 (ACI Feb. 94, Document), op.cit.

[74] Ibid.

[75] Point 9.2, pp.12-13, in ibid.

[76] Ground-handling became a core activity of German airports as a result of the post World War 2 situation. Initially West Germany was prevented from establishing an airline, and thus the loading / unloading of baggage was the only activity that France / US / UK allowed the German communities to undertake (interview: Bervd Nierobisch, Executive Director), ADV, Stuttgart, 10.7.97.

[77] Section 4, p.C.6, in ibid.

[78] Section 4, pp.C.12-C.14, in ibid.

[79] p.C.13, in ibid.

[80] 'Expanding Horizons. A report by the Comité des Sages for Air Transport to the European Commission' (Jan. 94.), p.5. Members of the Committee were as follows: Chairman: Herman De Croo, former Belgian Minister of Transport; Pieter Bouw, President of KLM Royal Dutch Airlines; Bjarne Hansen, President of Maersk Air (privately owned Danish airline, operating regional air services, although not a member of ERA); Geoffrey Lipman, President of the World Travel and Tourism Council; Henri Martre, Member of the Board and former Executive Chairman of Aérospatiale; Joao-Maria Oliveira-Martins, former Portuguese Minister of Transport; Gonzalo Pascual, Vice Chairman of Spanair (Spanish charter airline); Manfred Schölch, Vice Chairman of the Executive Board of Frankfurt Airport; Guillermo Serrano, Chairman of the Board of Amadeus (largest Computer Reservation System); René Valladon, Chairman of the Joint Civil Aviation Council (French transport union, Force Ouvrière); Jürgen Weber, Chairman of the Executive Board of Lufthansa German Airlines (source: 'Expanding Horizons', op.cit, p.54). Additionally, the support staff solely consisted of two members of DGVII; these were the Principal Administrator, Eckard Seebohm, and a further administrator from DGVII, Christopher Ross (source: 'Expanding Horizons', op.cit, p.54).

The purpose of the Committee, according to the report's chairman was to 'reflect on the future of aviation in Europe as an essential tool for economic and social development'. The report found, according to its Chairman, that the European airline industry was 'at a cross road'. The purpose of the Committee was to suggest 'political remedies' to this situation. The Chairman also stated:

> For six months it (the Committee) has listened to the often contradictory views of many knowledgeable people. It has contacted organisations, studied the many problems facing air transport, analysed data and weighted alternatives. The Comité has reached a broad consensus both on causes and remedies.
> ('Expanding Horizons', op.cit., p.5.)

However, the very first footnote of the report, immediately after the word 'remedies', reads: Two dissenting opinions were expressed: Messrs Schölch and Valladon on the issue of ground-handling services at Community airports and Mr Valladon on social issues. (Ibid, p.5.) This clearly displays the sensitivity of the issue of ground-handling, put forward by the Vice Chairman of the Executive Board of Frankfurt Airport and the Chairman of the French union, Force Ouvrière.

81 Interview: Bervd Nierobisch (Executive Director), ADV, Stuttgart, 10.7.97.

82 'Expanding Horizons', op.cit., p.38.

83 Note the emphasis on infrastructure, rather than commercial enterprise.

84 'Expanding Horizons', op.cit., p.38.

85 ACI claim that who operates the landside-handling is of 'little concern to the passengers' (see f./n. No.7, p.21, in ibid). However, a counter to this argument is that passengers, although not concerned with *who* is supplying the service, are concerned with the quality of service received, in addition to the price that they are charged for their ticket (a price directly affected by the cost of the ground-handling). Subsequently, most airlines claim that they wish to ensure that they have control over this service so that the quality remains high, and the cost low.

86 Point 4.2, p.6, in ibid.

87 As noted above, landside liberalisation is largely a non contentious issue. For example ADV (credited with being the prime influence in forming the 'non liberal' position of ACI) represents airports which have a liberalised landside handling market.

88 COM (94) 590 final, op.cit.

89 See: Trevor Soames (Partner, Norton Rose), 'Ground-handling...', op.cit.

90 Although in 1996 the Virgin Group, which owns Virgin Atlantic, established 'Virgin Express' (upon the purchase of 'Euro Belgian Airlines'), which operates a 'low cost' service throughout Europe.

91 Interview: Barry Humphreys (Director, Government and External Affairs), Virgin Atlantic, Crawley, 3.9.96.

92 Interview: Servisair, Manchester, op.cit. Also: John Willis (Chief Executive), Servisair, (panel discussion at Royal Aeronautical Society, Slots and Access to Airport Facilities, Conference), London, 21.1.97.

93 Servisair, 'Annual Reports and Accounts', 1996, pp.8-9.

94 Interview: Servisair, Manchester, op.cit.

[95] The view of Servisair in this regard is identical to that of British Airways. Thus, the UK government support was achieved not by the lobbying of Servisair, but by the lobbying of British Airways. Interview (via telephone), Rene Smith (Air Transport Section), UK Department of Transport, London, 16.10.96.

[96] Interview: Servisair, Manchester, op.cit.

[97] Ibid

[98] Source: ACI Europe.

[99] Interview: Mike Berman (Special Projects Manager), Business Air, Manchester, 14.1.97.

[100] Source: ACI Europe.

[101] Interview: Frank Pullman (Chief Executive), London Luton Airport, Luton, 14.1.97. Note, the passenger figure is the total number of passengers to pass through an airport in a given year, whether the airport is their departure point / destination, or they are in transit. The figure does not include cargo. See above for the differing definition of 'Traffic Unit'.

[102] Chris Gresser, 'Easy Jet determined to call the slots at Luton airport', *Financial Times*, 13.8.97, p.17.

[103] See Chapter 3, and O.J. C 339/41, pp.41-44, 31.12.91.

[104] 'Council Regulation (EEC) No. 95/93 of 18 January 1993, on common rules for the allocation of slots at Community airports', O.J. L 14/1, 22.1.93, Article 10.3, p.5. Seasons are divided into two per annum under International Air Transport Association (IATA) Guidelines. See: IATA, *Scheduling Procedures Guide* (Geneva: IATA, 1994).

[105] Point 3.4.4.1, p.9, in ibid.

[106] Interview: Peter Morrisore OBE (Managing Director), Airport Co-ordination Ltd., Heathrow, 10.1.97.

[107] Federal Regulation, 52, 180; 52, 184; 52, 199 (1985) cited in Eileen M.Gleimer, 'Slot Regulation at High Density Airports: How Did We Get Here and Where Are We Going?', in *Journal of Air Law and Commerce*, Vol. 61, No. 4 (May-June 1996), p.890. See also: David Starkie, 'Developments in Transport Policy- The US Market in Airport Slots', in *Journal of Transport Economics And Policy*, Vol. XXVIII, No.3 (Sep,1994), 325-329.

[108] Chris Allen (Head of Competition and Industry Affairs), British Airways, paper presented at IBC, 'European Air Transport Services and Airports' conference, Brussels, 6-7.2.97.

[109] Jonathan Faull (Director of Policy and Co-ordination, DGIV) and Frederick Sorensen, (Head of Air Transport Unit, DGVII), evidence to House of Commons Transport Committee, 15.7.96.

[110] See Chapter 6, Section 1.
Interview: Corinne Dussart-Lefret (Senior administrator, with responsibility for transport), DGIV, Brussels, 10.4.97.

[111] Interview: AEA, Brussels, op.cit.

[112] Kees Veenstra, 'Discrimination Under EC Slot-Allocation Rules In Favour Of New Entrants At Congested Hub Airports', paper presented at the Royal Aeronautical Society, 'Slots and Access to Airport Facilities', conference, London, 21.1.97.

[113] Interview: AEA, Brussels, op.cit.

[114] Interview: British Airways, op.cit; Chris Allen (Head of Competition and Industry Affairs), 'Comments on slot-allocation'; paper presented at IBC European Air Transport Services and Airports Conference, Brussels, 6-7.2.97; Chris Allen, 'Recent developments in slot-allocation', paper presented at European Air Law Association, annual conference, Copenhagen, 8.11.96; Robert Ayling (Chief Executive), 'The Shape of the Airline Industry in the 21st Century', paper presented at the 'FT World Aerospace and Air Transport Conference', London, 29-30.8.96.

[115] Interview: Corinne Dussart-Lefret, Brussels, op.cit.

[116] Interview: DGIV source of significant authority.

[117] Michael Skapinker, 'Airport slot trade may open up', *Financial Times*, 4.10.96, p.1.

[118] Chris Castles, 'The Application and Possible Modification of Council Regulation 95/93 on Common Rules for the Allocation of Slots at Community Airports', Final Report, Coopers and Lybrand, 17.10.95.

[119] Interview: Chris Castles (partner), Coopers and Lybrand (via telephone), London, 16.6.97.

[120] Point 2.57, p.18, in, 'The Application...' op.cit. Supported by Chris Castles (point 68, p.14), 'Development of airport slot-allocation regulation in the European Community', paper presented at IBC, European Air Transport Services and Airports conference, Brussels, 6-7.2.97.

[121] The airline Cargolux was not an AEA member at that time.

[122] Annex M, in, 'The Application...', op.cit.

[123] ERA, 'Summary of Key Issues Presented to Right Hon. Neil Kinnock, EU Commissioner for Transport', April, 1996.

[124] Chris Castles, 'The Application...', op.cit., Point 2.57, p.19.
The report also consulted Virgin Atlantic; however, as examined in Chapter 6, the report came out against Virgin's views on grandfather rights. See also Castles, 'Development...', Point. 74, p.15, op.cit.

[125] Point, 5, p.2 (Explanatory Memorandum),'Council Regulation (EEC) No... amending Council Regulation 95/93 of 18 January 1993 on common rules for the allocation of slots at Community airports' ('leaked' by fax, by officials within DGVII, to the media and interested parties, on, approximately, 1.10.96).

[126] Chapter 6, Section 2.

[127] Michael Niejahr (Administrator, DGVII), 'Recent developments in slot-allocation', paper presented at the European Air Law Association, Copenhagen, 8.11.96.

128 See Chapter 6. See also, various news reports, for example, 'Brussels warns of legal action over BA alliance', *Independent,* 14.1.97, p.14.

129 This statement is backed by many sources. Amongst them are the airlines themselves who could not state publicly that they have engaged in such an activity for fear of Commission action against them. However, the sources below confirm on the record that the activity takes place: (i) Interview: Philip Lowe (Chef de Cabinet to Transport Commissioner Kinnock), Brussels, 17.4.97. (ii) Interview: Peter Morrisore OBE (Managing Director), Airport Co-ordination Ltd., Heathrow, 10.1.97. (iii) Jonathan Faull (Head of Unit, DGIV), Frederick Sorensen (Head of Air Transport Unit, DGVII), evidence to House of Commons Transport Committee, 15.7.96. (iv) David Starkie (advisor to DGVII on slot-allocation, and adviser on the original Regulation on slot-allocation, Regulation 95/93), 'Recent Developments in Slot Allocation: What Can We Learn From The US Market In Airport Slots?', paper presented at the European Air Law Association, annual conference, Copenhagen, 8.11.96.

130 Interview: Corinne Dussart-Lefret (Senior administrator, with responsibility for transport), DGIV, Brussels, 10.4.97.

131 Interview: Andrew Clarke (Air Transport Policy Manager), ERA, Chobham, 23.1.97; ERA, 'Summary of Key Issues...', op.cit.

132 Interview: Malcolm Hart (Chief Executive), Gill Airways, Newcastle, 9.3.97.

133 ERA, 'Summary of Key Issues...', op.cit.

134 There is no precise statistic to draw this comparison. However, BA hold 36.5% of slots at Heathrow, its base airport. Lufthansa possess approximately 60% of the slots at Frankfurt, its base airport. It has been stated by an advisor to the Commission that the combined number of slots of all regional airlines at Heathrow presently equals a number less than the 'independent' carrier Virgin Atlantic - Virgin's figure is 1.9%. Therefore, this comparison partly displays the difference in the number of slots held. Source: David Starkie, 'Allocating Airport Slots: A Role for the Market?', paper presented at the Institute of Economic Affairs, London, 12.2.97.

135 ERA, 'Summary of Key Issues...', op.cit.

136 Interview: ERA, Chobham, op.cit.

137 Ibid.

138 ACI Europe, 'ACI Europe Initial Response Concerning The Potential For Leasing Slots To Airlines', 28.3.96.

139 Interview: Sheena Gibson (Head of European Policy Unit), ACI, Brussels, 10.4.97. Furthermore, although ACI clearly made its 'presence felt' regarding the Comité des Sages view on ground-handling, the Committee is strangely silent on the matter of grandfather rights and secondary trading. The Committee's only recommendation on the matter is that the 'European Commission must act to increase overall airport slot capacity' (Expanding Horizons, op.cit, p.20). However, this view is unanimously shared by all interests; the problem lies in how the Commission can go about this.

140 See: Chapter 6, and Scott Crosby (Partner, Kemmler Rapp Böhlke and Crosby), 'Recent Developments in Slot Allocation', paper presented at the European Air Law Association, Copenhagen, 8.11.96, p.2.

141 Hugh O'Donovan (Partner, Wilde Sapte, UK), 'Dealing and Trading in Slots', paper presented at the Royal Aeronautical Society, Slots and Access to Airport Facilities, Conference, London, 21.1.97.

142 Interview: ACI, Brussels, op.cit.

143 ACI Europe, 'Response To The European Commission's Discussion Document On The Revision of Council Regulation 95/93, May 1996; ACI Europe, 'Initial Response Concerning The Potential For Leasing Slots To Airlines', 28.3.96.; ACI Europe, 'Common Rules For The Allocation Of Slots At Community Airports - Review Of And Suggested Revisions to Council Regulation 95/93 of 18 January 1993', 24.11.95.; ACI Europe, 'ACI Europe Resolution On Slot Allocation', 13.11.92.; ACI Europe, 'ACI Europe Position Paper On Slot Allocation', 19.2.92.

144 ACI Europe, 'Response To...', op.cit.

145 ACI Europe, 'ACI Europe Position Paper...', op.cit.

146 Interview: Bervd Nierobisch (Executive Director), ADV, Stuttgart, 10.7.97.

147 Interview: Alistair McDermid (Head of Airport Strategy), BAA, London, 23.7.97.

148 Interview: Frank Pullman (Chief Executive Officer), Luton Airport, Luton, 6.1.97.

149 Interview: BAA, London, op.cit.

150 Interview: ADV, Stuttgart, op.cit.

151 Interview: Chris Walkden (Group Regulatory Affairs and Financial Planning Department), Manchester Airport, Manchester, 14.1.97.

152 House of Commons Transport Select Committee, enquiry into regional air services, 3.6.98.

153 Many airports are owned by local authorities. It is therefore in their interests to attract as many new airlines as possible.

154 For example, airports pay upwards of £1000 per delegate to attend the 'Routes' conference each year (a commercial conference organised by Airport Strategy and Marketing), where airports attempt to 'sell themselves' to airlines. Additionally, airports are willing to financially discount new entrants to their airports in order to win customers. See: Dinos Kyrou, 'Route support- a paper for Airport Strategy and Marketing, 27.1.98.

155 Source: Interview: Dr Barry Humphreys (Director, Government and External Affairs), Virgin Atlantic, Crawley, 3.9.96.

156 Source: David Starkie, 'Allocating Airport Slots: A Role...', op.cit.

157 Virgin Atlantic, 'A New Approach To Slot Allocation', March 1997.

158 Virgin Atlantic, 'Briefing Note - Current Air Transport Issues', March 1997.

Both documents were forwarded to DGVII of the Commission.

[159] Hugh Welburn (Economic Adviser to Virgin Atlantic), Institute of Economic Affairs, London, 12.2.97.

[160] Virgin Atlantic, 'A New Approach...', op.cit.

[161] Hugh Welburn (Economic Adviser to Virgin Atlantic), op.cit.

[162] Interview: Servisair, Manchester, op.cit.

[163] For information regarding inter-firm bargaining, see: John Stopford, Susan Strange and John S.Henley, *Rival states, rival firms - Competition for world market shares* (Cambridge: Cambridge University Press, 1992).

6 Commission Response to the Interest Representation of the Air Transport Industry

Introduction

Having examined the positions of the various groups in Chapter 5, this Chapter will now examine the application of their demands onto the policy formulation activities of the Commission, whilst focusing on the two hypotheses stipulated in the Introduction to this book. Hypothesis number 1 stipulates that there exists a reflection of the properties of liberal intergovernmentalism in the Commission - a *neo-liberal intergovernmentalism;* hypothesis number 2 stipulates that a *group cohesion* is a necessary pre-requisite in order to achieve successful interest representation. Furthermore, hypothesis number 2 stipulates that, in order to avoid a lowest common denominator of interest representation it is the large firms (or the elites), within the group which must be preferentially accommodated in order to ensure group cohesion and ultimately its successful representation.

The Chapter is divided into four sections. Section 1 will examine the background to the case studies. Section 2 will examine the positions of the legislative arm of the Commission for air transport, DGVII, and the *quasi-judicial* arm, DGIV, on the issues of ground-handling and slot allocation. Section 3 will compare the views of the Directorates which derived from the interaction with the elite and non-elite groups examined in Chapter 5. Section 4 will test the above hypotheses and will subsequently test the ability of DGIV to 'circumnavigate' the demands of elite interests.

Section 1

Ground-handling - Background

As previously discussed in Chapter 5, the European Commission stated that it received official complaints from the airlines British Airways, KLM, Air France, Lufthansa, Alitalia and SAS, concerning the alleged anti-competitive behaviour of ground-handling practices at Athens, Rome, Milan, Barcelona, Madrid, Lisbon, and Frankfurt airports. Frankfurt Airport was seen by DGIV as the key example of an airport which had not provided ground-handling liberalisation.[1] Airlines wished to reduce the cost of ground-handling services; the Gulf War of 1991, in particular, had led to a sharp reduction in the profitability of airlines resulting from increased oil prices and decreased levels of general travel. Ground-handling was seen as an area where the Commission could take action in order to reduce costs. At the time a spokesperson for KLM stated that if all the EU airports charged the same as Schiphol Airport, Amsterdam (an airport with a high degree of liberalisation), then KLM would save roughly £8.7m a year in ground-handling charges.[2]

Slot allocation – Background and Legal Opinions

As stated in Chapter 3, grandfather rights are practised globally and are recognised by Regulation 95/93. There exists no legal challenge to this recognition of grandfather rights by Regulation 95/93.

However, despite the non disputed legal legitimacy of grandfather rights, DGIV take the view that, in principle, the practice prevents air transport competition within the EU internal market.[3] Despite this, both DGIV and DGVII seem not to be convinced that grandfather rights are in themselves an abuse of a dominant market position, contrary to Article 86 of the Treaty of Rome.[4] This legal situation proved to be crucial in determining the options and strategies available to DGIV in pursuing its policy preferences.[5]

Legal opinion is divided as to whether the secondary trading of slots is lawful under Regulation 95/93. The view of both DGIV and DGVII is that, under present EC law, the secondary trading of slots is unlawful, as the practice is not clearly recognised by Regulation 95/93.[6] This view is supported by Crosby;

> There is an assumption that those possessing slots may trade them and freely make money out of them. No thought seems to be given to the question of the ownership of the slots...
>
> The question of slot ownership has never been resolved in Europe. Slots are 'in the possession of airlines' and discussion goes no further. In fact, current EC slot legislation is not based on any firm property law principles, but on mere assumption.[7]

Crosby continues that airlines pay nothing initially to have a slot allocated to them; an economic anomaly where a valuable asset is made available gratuitously. Furthermore, the airport operator who built the infrastructure that allows the slot to exist (i.e. the runway and support facilities, such as the airport terminal), has no rights over it; a second economic anomaly. Property rights are not created this way. Rather, they accrue to the person who has created the asset by their effort or investment; the founding principle of all property law, he concludes.[8]

Robert Webb states that Regulation 95/93 makes it a criminal offence to deal with slots in any way other than that set out in the Regulation. However, Regulation 95/93 does not make the practice of secondary trading a criminal offence itself; he argues:

> A criminal offence requires the forbidden conduct to be clearly and specifically forbidden. The benefits of any doubt must be given to the defendant and the defendant must usually have a *mens rea* - a 'guilty mind'.
>
> My view, as a trial lawyer and reader of documents is that slot trading is allowed, and the present 'grey market'[9] is, in actuality, lawful.[10]

O'Donovan queries the ambiguity of the relevant article of Regulation 95/93, where it states:

> Slots may be freely exchanged between carriers from one route to another, by mutual agreement, partial take-over, or unilaterally,

and subsequently argues that a 'great deal of confusion' arises from this particular provision.

> The provision completely fails to refer to monetary payment. Monetary payment could not be regarded as unlawful in the context of that provision.
>
> The UK Civil Aviation Authority has stated that secondary trading violates the spirit of the Regulation. The CAA argues that the Regulation 'envisages a cash free transfer of slots', when such transfers take place. However, the Regulation does no such thing.

As well as querying this legal ambiguity, O'Donovan questions why it was the case that such a significant section of Regulation 95/93 was written in this way.

> I have yet to find anyone who can explain why it has been drafted in this way. Even the Commission Discussion paper of April 1996 described the wording as 'too cryptic'.[11]

The Commission Discussion paper, to which O'Donovan refers by this statement[12] was in fact written not by the Commission as a whole, but by the Transport Directorate DGVII; splits between DGVII and DGIV on the issue were apparent during this time. However, divisions between the two Directorates were also apparent during the formulation of Regulation 95/93 concerning both secondary trading and grandfather rights.[13] Subsequently, it appears that the issue of secondary trading was left deliberately vague. As a result, Article 14 of the Regulation was the proviso allowing for a delay in the decision making process on these issues, due to the Commission's inability to reach a meaningful consensus;

> The Council shall decide on the continuation or revision of this Regulation [95/93] by 1 July 1997, on the basis of a Proposal from the Commission to be submitted no later than 1 January 1996.[14]

A situation presently exists where the legitimacy of grandfather rights and the illegality of secondary trading is not disputed between DGs IV and VII.

There exists, therefore, a consensus between DGIV and DGVII on the status of grandfather rights and secondary trading, as stipulated by Regulation 95/93.[15] However, although there exists a consensus within the Commission on the interpretation of Regulation 95/93, DGIV and DGVII hold diametrically opposite positions on the way in which reform should take place on these matters.

Section 2 - Aims of DGs IV and VII

Aim of DGIV – On ground-handling

Ground-handling is a matter which has been pursued by DGIV since it received the first airline complaints regarding the issue.

The initial press release, concerning the Commission's investigation into ground-handling monopolies, was issued solely by DGIV.[16] The Press Release makes clear the main concerns of the Competition Commissioner Van Miert, and the action which he deemed necessary. Van Miert stated that as the three packages of air transport liberalisation were now in place,[17] the Commission had to ensure that their effect was, 'not diluted by restrictions on competition in the provision of ground-handling services at airports.' Van Miert added that the complaints were,

> ...being examined by officials in order to deal with any abuses of dominant positions which might arise; I am convinced that the real problem lies with the granting to operators by the authorities of exclusive or special rights for the provision of ground-handling services. These monopolies restrict competition to the detriment either of airlines prepared to supply these services,[18] or of independent ground-handling companies...[19]
>
> So that action can be taken as swiftly as necessary it should in my opinion take the form of a Directive adopted by the Commission itself using the powers under Article 90 of the EEC Treaty.[20]

The legal justification for the issuing of a Commission Directive based on Article 90 of the EC Treaty is that, after investigation, the Commission must be satisfied that certain parties are abusing various competition criteria.[21]

Following the Commission Press Release, two high ranking officials within DGIV stated,

> ...it appears that the existence of ground-handling monopolies in favour of public or private undertakings based on special or exclusive rights is unlawful pursuant to Article 90(1) in combination with Article 86 of the EC Treaty..
>
> Article 90(1) provides that in the case of public undertakings and undertakings which have been granted special or exclusive rights, Member States shall neither enact or maintain in force any measure contrary to the rules of the EC Treaty...
>
> Article 86 declares that all behaviour of undertakings constituting an abuse of a dominant position, within the Common Market, or a substantial part thereof, which may affect trade between Member States is incompatible with the Common Market.[22]

For the Commission to issue an Article 90 Directive it would need to find that, '...there was no objective necessity for the existence of the monopoly for the provision of ground-handling services'. Crucially, the DGIV administrators added,

> It appears that a complete monopoly is rarely, if ever, objectively needed to enable ground-handling services to be carried out.[23]

DGIV therefore comprehensively dismissed the points subsequently made by ACI in its document released in February 1994;[24] this document was concerned with the 'limiting factors' to liberalisation at EU airports,[25] DGIV stated:

> Sometimes it is said that high investment costs, lack of space, safety and security requirements make it impossible to allow more than one company to provide ground-handling services. However, there are airports in the EC and elsewhere where multiple ground-handlers are routine and where neither security nor efficiency seems to suffer.

> As regards the argument of limited space, it should be noted that it is the number of flights serviced which determines the amount of space required for the accommodation of personnel and equipment and not the number of companies which provide ground-handling services.[26]

The DGIV officials supported the views put forward in the Commission Press Release, namely that an Article 90 Commission Directive would be their preferred option.

> The alternative option of a Council Directive[27] implies a somewhat more lengthy process involving formal consultations.
>
> A Commission Directive ...would have the advantage of the Commission being able to provide a relatively speedy solution to the existing core problems in this sector.[28]

A Commission Directive could potentially have created framework measures to accompany the three packages of air transport liberalisation already in place at that time. These measures would have set out certain minimum rules which would have had to have been implemented (under Community law) by Member States, airports and ground-handlers.[29] A Commission Directive (arguably far more effective than a Council Directive), was, therefore, the preferred option of the Competition Commissioner.[30]

View of DGVII – On Ground-handling

DGVII did not become deeply involved in the ground-handling situation in a legislative manner until it was required to formulate a Proposal for a Council Directive.[31] DGVII needed to be 'seen to be taking action' resulting from the ground-handling situation and the adverse publicity this produced for EU competence in air transport liberalisation.[32] However, the 1993 Commission Proposal for a Council Directive contradicted many of the views of DGIV. Fundamentally, the Proposal rejected the claims concerning ground-handling competition previously put forward by DGIV. Additionally, the potential reasons against liberalisation, originally rejected by DGIV, are used both *de jure* and *de facto* in the Commission Proposal, as potential legitimate justifications to delay or (in some areas), prevent the liberalisation of ground-handling.

The Commission Proposal for a Council Directive, drafted by DGVII, uses precisely the same reasons to delay or hold back liberalisation which were originally rejected by DGIV. The Commission Proposal subsequently presented a complete *volte face* on the relevant issues. Indeed, although the Explanatory Memorandum to the Proposal admits,

> The current organisation of the ground-handling market no longer meets the needs of air transport, nor does it comply with the principles of the single market,

the Memorandum goes onto state:

> The supply of ground-handling services is, however, subject to a number of practical constraints, which vary according to the type of service. These constraints concern in particular - available capacity and space...systems which cannot be duplicated, as in the case of baggage sorting...safety and security...
>
> adding,
>
> ...the Commission feels that access to the ground-handling market should be such as to allow users to choose the supplier who comes closest to their quality and cost requirements. However, access should also be adapted to the features of the various types of services, taking account of the existing constraints of social repercussions as well as the need to keep the airports operating efficiently.
>
> The Commission therefore favours the introduction of liberalisation as well as the accompanying measures.[33]

An analysis of the Proposal displays how DGVII 'watered down' the legislation in order, it seems, to appease dominant, elite, interests. The issues listed below are those which these dominant interests were attempting to pursue, and are summarised into 4 separate 'goals'.

Goal 1 The minimisation of the number of airports included in the legislation.

The fewer the number of airports to be included within the legislation, the fewer the number of airports that would have to adopt liberalising measures. Setting a 'high minimum threshold' for liberalisation would therefore result in a fewer number of airports being required by EC law to liberalise.

Goal 2 The maximisation of delay in the introduction of liberalisation.

The longer the delay in the enforcement of the legislation, the longer it would be before airport operators were legally required to liberalise.

Goal 3 To present 'barriers to entry' to potential competitors.

The greater the financial (or other) barriers to new entrant ground-handlers, the smaller the possibility of the airport operator facing effective competition.

Goal 4 To guarantee the ability for an airport to supply ground-handling services.

Various airports wished to be guaranteed the option to ground handle, without having to embark on any 'selection procedure' which might be required of other potential or existing ground-handlers.[34]

Achievement of Goals by DGVII in Proposal

Achievement of Goal number 1 in Proposal

Chapter 4 stipulated that those interests, such as ERA, wishing for as free a ground-handling market as possible also wished for the maximum number of airports to be included in the Directive to achieve this aim.[35] However, the Proposal stated that only airports whose annual passenger number was greater than two million passengers were covered by the proposed liberalisation to third party handling. Subsequently, a fewer number of airports were included in the Proposal.[36] This fact was a particular disadvantage to members of ERA who operate from a high number of the smaller airports within the European Union.

Achievement of Goal 2 in Proposal

The relevant Article and its Explanatory Memorandum state that where an airport has a 'specific situation' (safety, technical feasibility, space and capacity restraints), the relevant Member State may limit the number of third party, or self handlers to a single supplier, which may be the airport itself, for a limited period of time. This delay may be for a maximum of three years. However, the Article also stated that the Member State may grant a new exemption (duration of which is not stipulated), at the end of the three year period.[37]

Achievements of Goal 3 in Proposal

1. Centralised infrastructure and access to installations - Centralised infrastructure in ground-handling is described as: centralised baggage sorting, de-icing of aircraft, water purification and de-icing systems. Access to such systems is essential if one is to self-handle, or provide third party services. However, according to the Proposal the technical management of such systems may, if the Member State wishes, be reserved to the airport operator. Furthermore, the airport operator may make use of such infrastructures 'obligatory'.[38] For example, if an independent ground-handler wished to operate a separate water hydrant (an essential facility), the airport operator may prevent them from doing so. This fact may well hinder the operation of a self or third party handler if the airport wishes to prevent competitors from having access to these essential facilities.[39]

2. Definition of self-handling. The Proposal stipulated that self-handling is where an airport user,

> ...directly provides for himself one or more of the categories of ground-handling services and concludes no contract of any description with a third party for the provision of such services.[40]

Therefore an airline, operating a franchise[41] into a Community airport, would not gain the liberalisation advantages stipulated for self-handling, as the handling of a franchisee by its franchisor would not be considered a 'direct provision'.[42] This limited definition was also a particular disadvantage to ERA members, as many of the operations of regional carriers are operated under franchise. Subsequently, this definition provided less choice for regional airlines to choose their ground-handlers.
3. Selection of suppliers - The relevant Article stipulates that twelve months at the latest after the entry into force of the Directive, the Member States must establish a 'Users' Committee' at all airports which qualify for the liberalisation of third party ground-handling (i.e. all Community airports handling over 2 million passengers annually). The Users' Committee, to consist of all airlines that use a particular airport is to make decisions by a voting system weighted by the volume of traffic that a particular airline operates at the particular airport. The task of the Users' Committee is to select the third party ground-handler(s) which will be separate from the airport operator.[43] This method of decision making by the Users' Committee substantially disadvantages regional airlines as the larger airlines on the Committee will have a dominant voice, due to the decision making mechanism.[44] Furthermore, the Users' Committee cannot select a supplier for more than seven years.[45] This factor has been described as a substantial disincentive for anyone to operate such ground-handling services, as it is considered by some as not enough time to recover the capital expenditure incurred with market entry.[46]

Achievement of Goals 2 and 3 in Proposal

The Commission Proposal stated that all airports in the EU must allow self-handling within two years of the entry into force of the Directive.[47] However, the Proposal also stated that the Member States may limit the number of self-handlers to two; the Proposal therefore ends the possibility of legislation that all airlines will be able self handle, if they so chose. The relevant Article therefore both delays the introduction of liberalisation, (Goal 2), in addition to presenting a barrier to entry (Goal 3).

Achievement of Goal 4 in Proposal

The Proposal stated that within two years of the entry into force of the Directive, Member States must allow free access to the market for third party ground-handling services.[48] However, there were a multitude of caveats in order to prevent the realisation of this aim. Foremost amongst them was the fact that the relevant Article also allowed a Member State to limit the number of third party handlers to two. Although the relevant Article stipulated that at least one of the third parties may not be controlled directly or indirectly by the airport or the flag carrier airline,[49] the Proposal also stated that, where the number of third party handlers is limited to two by the airport, the airport itself has an automatic right to third party handle; conversely, the flag carrier has no such automatic right.[50] This situation severally disadvantages flag carriers who already third party handle for other airlines, as they may have their right to do so confiscated from them. The Article concerning third party handling therefore guarantees ground-handling to the airport (Goal 4), if the airport so wishes, regardless of any economic or commercial justification.

Clearly, the requirements of DGIV differed substantially from the Commission Proposal, formulated by DGVII. DGIV essentially stated that the reasons given against liberalisation were little more than 'excuses' to prevent liberalisation. Conversely, the Commission Proposal utilises these same 'excuses' as reasons to hold back and delay ground-handling liberalisation.[51]

On slot allocation - Positions of DGIV and DGVII on secondary trading and grandfather rights in the attempted re-formulation of Regulation 95/93

DGVII, the Transport Commissioner, and the members of his Cabinet, believe that the clear legal recognition of a secondary slot market is the correct path to follow for the revision of Regulation 95/93. DGVII believes that the current 'grey market' in slots should be clearly defined in favour of the practice. Furthermore, DGVII believes that grandfather rights should continue to be recognised in any new Regulation.[52]

Conversely, DGIV oppose the continued recognition of grandfather rights and believe that any new Regulation should clearly abolish such rights. DGIV favours a system where such rights are time-limited, as opposed to the indefinite system of grandfather rights which currently exists. DGIV also oppose the clear legalisation of a secondary slot market, arguing that secondary trading should be clearly outlawed in the revision of the Regulation.[53]

In April 1995 the auditors Coopers and Lybrand were commissioned to review possible ways in which Regulation 95/93 might be revised. Its report stated its support for grandfather rights, as well as the secondary trading of these grandfather rights.

> Some airlines have argued that combining grandfather rights with monetarised secondary trading would be appropriate for Europe's airports, and would be a more powerful and efficient means of facilitating new entry[54] than altering the administered rules by which new capacity is allocated...
>
> ...it appears that the economic incentive on airlines would be sharpened...as carriers rationalised their slot holdings in a manner that would allocate capacity according to individual airline's valuations of slots rather than historic rights.[55]

Subsequently, in a conference speech to the airline industry, the author of the report reiterated his support:

> ...the development of a secondary trading mechanism for slots at the congested European airports would be a useful step to address the weaknesses in the current administered system of slot allocation. It would introduce a strong monetary incentive to encourage the release of increasingly scarce airport capacity at congested airports by those airlines less able than others to make efficient use of slots.[56]

The report supports DGVII's view that what is currently a 'grey market', should be made a clear one, in order to 'free up' this market and allow the 'true valuations' of slots to be revealed.[57] However, the report, its brief, and the choice of its author, Coopers and Lybrand, were entirely the responsibility of DGVII;[58] conversely, DGIV played no part in the formulation of the report. Moreover, although the report states that Coopers and Lybrand were engaged by the Commission to compile the report,[59] the 'Commission' is used as a generic term to describe the actions which were, in actuality, taken by DGVII.

Following the publication of the Report an official from DGVII publicly stated: 'The Coopers and Lybrand Report will provide the basis of the new Proposal for a Regulation.'[60] However, a senior source within DGIV stated:

> What particularly struck DGIV was the lack of information in the Coopers and Lybrand report. It contains very little data to support its views. Its views on grandfather rights and secondary trading are based simply on opinion. However, although there is very little data, it clearly stated what it wished to happen- this we found surprising.[61]

Indications of 'friction' between DGIV and DGVII were intensified by the leaking of a 'draft' Regulation, which, in addition to recognising grandfather rights, also categorically permitted the monetarised secondary trading of slots between airlines. The leaked draft Regulation directly contradicted the known views of DGIV on the issues of grandfather rights and secondary trading. It stated:

> [The] confiscation of slots, either by time-limiting the grandfather rights, or by imposing the return to the slot pool of a certain percentage of the slots...potentially undermines the longer term competitiveness of the incumbent air carriers, particularly of the carriers having their home base at the congested airport in question. Also, it is not compatible with the established and internationally accepted system of slot allocation and may lead to serious repercussions for Community air carriers elsewhere in the world. For these reasons any forced surrender of slots is strongly resisted by air carriers and Member States.[62]

On secondary trading it stated:

> After careful consideration of all other options to stimulate the necessary flexibility at highly congested airports, the Commission has opted to facilitate exchanges of slots between air carriers by removing the obligation to give a slot in return for a slot and by allowing the monetarised trading of slots between air carriers.[63]

Therefore, the leaked 'Proposal' contained and indeed clarified the views on grandfather rights and secondary trading contained within the Coopers and Lybrand Report. The leaking of the draft Regulation by DGVII was the first public 'confrontation' with DGIV. This was followed by an open letter from Transport Commissioner Neil Kinnock to the *Financial Times* in April 1997, which stated: '...our proposals for a revised airports slots regulation will shortly be published...'[64] This view was previously supported by on the record comments in a speech by Neil Kinnock's Chef de Cabinet at a conference in February 1997, where he stated: '...the Commission is about to propose improvements to the 95/93 slot allocation Regulation.'[65] Furthermore, the Chef de Cabinet stated in interview in April 1997 that the Proposal was 'already formulated', and would become an official Commission Proposal in May 1997.[66] However, the 'Proposal' that DGVII intended to release (a 'Proposal' clearly legalising grandfather rights and secondary trading), was diametrically opposed by DGIV. By stating the imminence of the Proposal, DGVII seems to have attempted to 'force the hand' of DGIV into allowing DGVII to cast light on its particular views on the two disputed issues. By constantly stating that the Commission was 'on the verge of publishing the Proposal', DGVII seems to have been attempting to push DGIV into a situation that it could not retreat from. Publicly, DGVII has always used the terminology, 'the Commission', when speaking of the 'the Commission's' views on the disputed issues; however, DGVII seems to have been speaking of its own views. Moreover, personnel within DGIV not only opposed the views of DGVII, but deeply resented the fact that DGVII was leaking such information to the media. Following the deliberate leak of the draft Regulation, contradictory statements were simultaneously submitted by DGIV and DGVII. Indeed, a source at DGIV stated to the *Financial Times:* 'Selling slots is not something we can accept. It is not legal.'[67] Simultaneously, it seems possible that the Chef de Cabinet to Transport Commissioner Kinnock[68] in a communication with, *The Independent,* stated:

> Slot selling is an economic reality and it is done all the time. So are we going to try to fight against a reality or are we going to try to give a framework to this practise in order to avoid monopolies and guarantee fair competition?[69]

In February 1998, Competition Commissioner Van Miert stated publicly:

> I am personally convinced that secondary trading is not the appropriate solution. There are lots of reasons for this, and I will mention only two of them...
>
> Some people say that trading is the best solution because the slots will be bought by the most efficient airlines. Is this necessarily so? What about smaller carriers which do not have deep pockets? Should one increase the obstacles to market access for them?
>
> The economic analysis is based on the assumption that there is a market for slots. [However] a market requires buyers and sellers. In congested airports like Heathrow we know the buyers, but who are the sellers? Maybe as a first step, some small airlines will need money and would sell some slots, but after that? The likelihood is that slots would be sold and bought only once, for the benefit of a few already dominant carriers.
>
> The second reason to reject the trading of slots is the US experience. As I mentioned, the United States authorised the trading of slots for internal services, and yet, what is the result? Dominant companies bought the slots and reinforced their dominant positions.[70]

DGIV has the support of several of the MEPs on the European Parliament Transport and Tourism Committee, including its former rapporteur;

> I prefer the idea of limiting slots for a period of time, rather than possessing indefinite grandfather rights. Grandfather rights make it very difficult for smaller airlines to get into airports.[71]

This position receives robust support from member of the Transport and Tourism Committee, Brian Simpson, MEP, who states: 'Can you have grandfather rights or slot trading in a liberalised air transport system?'[72]

Section 3 - A comparison of the positions of DGs IV and VII to the examined interests, and the interaction of interest groups in relation to the Directorates

DGVII – The response to interest representation

Ground-handling

On ground-handling, Airports Council International clearly had the most effective role, firstly in preventing a Commission Directive based on Article 90 (thereby apparently scuppering the chances of any substantial involvement by DGIV), and subsequently by ensuring that the Proposal for a Commission Directive (formulated by DGVII), was weakened in the ways outlined above. These weaknesses were indeed carried through to the final Council Directive.[73]

If one examines the position of the European Regional Airlines Association, it is clear that its demands were clearly not met by the Proposal, even though the original wishes of DGIV for a Commission Directive would have addressed ERA's concerns.

The most intriguing group on ground-handling is the Association of European Airlines. As has been stated throughout this book, the AEA consists of elite interests; yet the views of the AEA are barely reflected in the outcome of the Commission Proposal. What does such a predicament display to an analyst of this situation?

Application of hypothesis number 1

Does this position of DGVII display a reflection of a form of liberal intergovernmentalism within the Commission - a *neo-liberal intergovernmentalism?* Clearly, the wishes of the non-elite group, the ERA, were not reflected in the Commission Proposal. DGVII when drafting the Proposal therefore had 'a choice of elites'- the airports and the flag carrier airlines. As stated in Chapter 4, the elite airlines present an extremely powerful lobbying group as the AEA consists primarily of national champion interests.[74] However, although the view of DGVII more closely reflects the view of the airports, I hold that the response of DGVII was indeed a reflection of liberal intergovernmentalism - a *neo-liberal intergovernmentalism.* I validate this claim for two reasons.

1. Only three flag carrier airlines (BA/KLM/SAS) had the full support of their representative governments in their initial complaints on ground-handling. The view of the Air France Chairman was not reflected by the French government, and he was eventually dismissed from his post by the French Transport Ministry.[75] Moreover, the views of Alitalia and Lufthansa (also the original complainants to DGIV) were not reflected by the Member States Italy and Germany, respectively. Furthermore, and crucially, the views of the German Airports ADV and the German government, together with the French Air transport unions with the support of the French government, opposed liberalisation. The role of ADV in this area cannot be overstated. Indeed, the Executive Director of ADV has stated:

> ....Germany is one of the most important countries in the EU; the Commission did not want to have a confrontation with the German government, and run the risk of the whole Directive failing. The Commission therefore made sure the final Proposal was more 'flexible'...[76]

2. One of the key determinants of 'eliteness' as specified in Chapter 1, is that elites normally employ the highest number of people in that particular sector of industry. In head-on terms, airports employ more personnel than airlines. An airport produces approximately 8,000 indirectly related, and 2000 directly related jobs for each 1 million passengers that passes through the airport.[77] A sample comparison of this measure of eliteness can be seen in the table below.

Table 6.1 A table to show a comparison of employment between a sample of flag carrier airlines and the airports in their home state

Airline	**Air France**	**Alitalia**	**Lufthansa**
No. of employees (Direct)	37,329	18,385	32,240
	French Airports	**Italian Airports**	**German Airports**
No. of employees (Indirect)	824,000*	445,000*	968,000*
(Direct)	206,000*	111,250*	242,000*

*To nearest 1,000
Sources:
AEA 'Yearbook' 1996, pp.40-54.
ACI Europe, 1992 data (calculation based on total number of passengers passing through ACI member airports in France, Italy, and Germany).

In such a 'head on', albeit crude comparison of eliteness, airports clearly win hands-down. Therefore, the answer to hypothesis 1 that there exists a *neo-liberal intergovernmentalism* in that the properties of liberal intergovernmentalism are indeed reflected within DGVII. However, *eliteness is relative a concept;* the 'level of eliteness' depends upon which interests are competing for the ear of the policy formulator. Both the AEA and ACI consist of elite interests, the former primarily, and the latter as a non exclusionary group. However, as policy formulators DGVII acquiesced to the demands of the elites within this non exclusionary group. Indeed, the Deputy Secretary General of the AEA stated:

> You may have the correct argument, but if there are particular factors against you, like unemployment, you have no chance.[78]

Application of hypothesis number 2

As examined in Chapter 5, ACI formulated a highly detailed position on the issue of ground-handling. Fundamentally, it produced a group cohesion, as stipulated as necessary by hypothesis 2. The position of ACI subsequently reflected the group's 'most demanding interest' ADV - the German Airports group.

Together with the pressure applied nationally by ADV in combination with the union ÖTV, as well as the Bundasrat, ADV was able to lobby most effectively throughout this episode. This is not to downplay the role of the French Union *Force Ouvrière*; its support was crucial in the Dissenting Opinion in the Comité des Sages document.[79] It prevented the appearance that the Dissenting Opinion was a solo view; indeed, the initial Commission Proposal for a Council Directive stated that 'some members expressed dissenting views.' This phraseology is, arguably, infinitely more effective than, 'a member expressed a dissenting view.' Therefore, the fundamental difference between ACI and AEA was that ACI produced a group cohesion, and the elites within the group (in particular ADV), were able to dominate the decision making process, as stipulated in hypothesis number 2.

This situation would seem to support the Greenwood hypothesis that groups must weigh the interests of the large players against the small, in order for their interest representation to be effective; 'Without the scales being tipped in favour of the large players, a federation is doomed.'[80] Furthermore, as stipulated by McLaughlin, Jordan and Maloney[81] ACI realised that the diverse tactics of large firms can undermine the representation of groups. Therefore, rather than have ADV establishing such diverse tactics, ACI acquiesced to ADV's 'minimalist' demands.

Conversely, the AEA were in a situation where interests within the group (Iberia, Tap and Olympic), acted independently, thus breaking away from the group representation. The group representation in itself was subsequently inadequate and was a lowest common denominator view. However, the relevance of lowest common denominator decision making in groups brings an extra dynamic to the hypothesis that a lowest common denominator produces a weaker presentation for an interest group than a position which is formulated by a form of majority decision making. Both ACI and AEA produced lowest common denominator positions, yet the lowest common denominator position produced by ACI was substantially more effective than the lowest common denominator produced by the AEA. An analyst of this situation must enquire as to why this was the situation.

As stipulated in Chapter 5, the position produced by ACI, although a lowest common denominator, was not a position which adversely affected the other interests within the group.[82] Furthermore, although a lowest common denominator position was produced, interests were not willing to 'break from the pack' and separately represent themselves independently to policy formulators. Conversely, the lowest common denominator produced by the AEA created a high level of dispute between its composite interests. Subsequently, interests pursued their separate paths, thus undermining the representation of the group as a whole. Therefore, one can ascertain the following. Whether or not an interest group produces its position by lowest common denominator or by a form of majority vote is largely irrelevant, what is most important is the net result of the decision making, i.e. if a firm within a group 'breaks from the pack' then, according to Jordan and McLaughlin, 'Groups...become paralysed because they cannot maintain consensus amongst conflicting interests'.[83] This was the fundamental difference between the position of ACI and AEA on ground-handling. The AEA achieved such a paralysis, as the AEA as a group was unable to maintain any consensus amongst its conflicting interests. Conversely, ACI achieved a consensus which airports were happy not to circumnavigate. Therefore, in regards to hypothesis number 2, *group cohesion* was indeed necessary in order to achieve ACI's position. Furthermore, the elites within the group (particularly the German elite, ADV) were allowed to dominate. Within the AEA this did not happen. However, a lowest common denominator position does not in itself lead to weak representation of the group; both AEA and ACI produced lowest common denominator opinions; it is the outcome of the lowest common denominator (i.e. whether the group remains coherent or becomes split), which may ultimately determine the effectiveness of the group.

Slot allocation

A key difference between the group representation in the attempted revision of the slot allocation Regulation and the formulation of the Proposal for the ground-handling Directive is the near non participation of ACI in the former. Therefore, there is no 'choice of elites' for DGVII, a choice which existed in the formulation of the Proposal for a ground-handling Directive. ERA and the AEA share their views on grandfather rights and only the firm Virgin Atlantic opposed these views.

Application of hypothesis number 1

A *neo-liberal intergovernmentalism* exists inasmuch as the views of DGVII reflect the views of the elite interests, which are, in this case, the airlines of the AEA. However, as there exists no 'choice' to make between the eliteness of interests, the only interest publicly opposing the views of the AEA is Virgin Atlantic; itself not a member of a relevant interest group in this situation. Therefore, a situation exists where the views of all the interests, bar one, were reflected by the views of DGVII. This is subsequently a reflection of *neo-liberal intergovernmentalism,* inasmuch as it reflects a near unanimity of interest representation, although it does not create a situation of preference of an elite group over a non-elite group. Having established this fact, it may well be the case that the non-elite group the ERA, realised that to oppose secondary trading would indeed create a situation where it was placed 'head on' against the elite AEA. The fact that ERA appears to allow an elite domination within itself (as stipulated in hypothesis number 2), may well be reflective of a realisation by such groups of the dominance of elite over non-elite interests in their representation to policy formulators.

Application of hypothesis number 2

Despite the fact that the members of the AEA are unwilling to admit in public their support for secondary trading (except for British Airways), a clear unanimity appears to exist across the AEA in this area. Therefore, in response to hypothesis number 2, a group cohesion exists across the AEA and no lowest common denominator has been established.[84] As stated in Chapters 4 and 5, in order to bolster the representation of ERA, the air transport policy manager seems to have acquiesced to the demands of the airlines within ERA which are in themselves subsidiaries of elite airline interests. Therefore, in order to achieve more effective interest representation, the air transport policy manager appears to have attempted to implement the Greenwood hypothesis within his non-elite group. However, as an inevitable result of this practice, the view of ERA will ordinarily reflect the view of the AEA. Therefore, the views of ERA are likely, *de facto*, to be the views of the AEA. Although this practice may well create a situation where it would appear that the views of ERA are being reflected by policy formulators, the relevance of ERA as a lobbying organisation may well be undermined. Member airlines which subscribe to ERA may eventually realise that it might be more cost effective to 'free ride' the representation of the AEA if ERA is ultimately going to follow the views of the AEA.[85] Subsequently, application of the Greenwood hypothesis in a group which is in itself non-elite may well lead to the whole necessity of the group being brought into question by those interests which subscribe to its membership.

DGIV – The response to interest representation

Ground-handling

As examined comprehensively above, DGIV dismissed the views put forward by 'anti-liberalising' elements, in particular ACI; DGIV also wished to achieve substantially more in the way of liberalisation than the 'watered down views' of the AEA. What does this situation tell the analyst about the examined hypotheses?

Application of hypothesis number 1

It probably cannot be argued that a reflection of liberal intergovernmentalism - a *neo-liberal intergovernmentalism* - existed in relation to the views of DGIV and elite interests in this area. Indeed, the views of DGIV more closely reflected the views of the non-elite interest group, the ERA. Therefore, the view of this *quasi -judicial* element of the Commission was substantially different from the legislative element - DGVII.

Application of hypothesis number 2

Hypothesis number 2 proves to be somewhat irrelevant in the case of this *quasi-judicial* institution. As argued extensively in Chapter 5, ACI achieved group cohesion on this issue; yet its views were diametrically opposed by DGIV. Hypotheses 1 and 2 therefore cannot be applied to the position of DGIV in the way in which they can be applied to the position of DGVII in ground-handling. However, hypotheses 1 and 2 played a substantial role in determining the *path* that DGIV must take in order to achieve its goals. Neither hypotheses can be applied to the *position* that DGIV established on ground-handling. However, the *path* which DGIV needed to follow in order to achieve its position did indeed alter, as a result of a *neo-liberal intergovernmentalism*, and as a result of the effective lobbying as a result of group cohesion. Further examination of this 'third hypothesis' is contained in Section 4 of this Chapter.

On slot allocation

Application of hypothesis number 1

With the exception of Virgin Atlantic, none of the views of interests, either elites or non-elites, were reflected by DGIV. In stark contrast to the situation which existed within DGVII, the view of DGIV was clearly not a *neo liberal intergovernmentalist* situation, as the position of DGIV was not an elite preference over non-elite interests.

Application of hypothesis number 2

Both of the groups in this case study, the AEA and the ERA, presented positions on the issues of secondary trading and grandfather rights which reflected group cohesion within their respective groups. In particular, ERA seems to have allowed the elites within the group to dominate, a practice stipulated as vital by Greenwood for successful interest representation. However, their views were not reflected by DGIV. Subsequently, hypothesis number 2, the importance of domination by elites within the group (stipulated by Greenwood),[86] and the importance of group cohesion (stipulated by Jordan, McLaughlin and Maloney),[87] may become, when determining the position of the *quasi-judicial* element of the European Commission, largely irrelevant.

Below are two tables to summarise the positions of DGs IV and VII as well as the examined interests on the issues of ground-handling and slot allocation.

Table 6.2 A table to compare the interests and the positions of DGs IV and VII on the issue of ground-handling

Non-elite		
ERA	**Virgin**	**Servisair**
Fully supported liberalisation	Non active	Fully supported liberalisation
Elite		
AEA		
Supported liberalisation with reservations		
Non exclusionary		
ACI		
Against liberalisation		
Directorate Generals		
DGIV		
Fully supported liberalisation		
DGVII		
Formulated Commission Proposal which substantially 'watered down' the aims of DGIV		

Table 6.3
A table to compare the interests and the positions of DGs IV and VII on the issue of slot allocation

Non-elite		
ERA Supports grandfather rights and 'not opposed' to secondary trading	**Virgin** Against grandfather rights and secondary trading	**Servisair** Inactive
Elite		
AEA Supports grandfather rights and secondary trading		
Non exclusionary		
ACI Inactive		
Directorate Generals		
DGIV Against grandfather rights and secondary trading		
DGVII In favour of grandfather rights and secondary trading		

Section 4 - 'To defy the hypotheses?' - the actions of DGIV in the case studies

Ground-handling

Clearly, the role of the two hypotheses cannot be applied equally to a legislative arm of the Commission, and the *quasi-judicial* arm. Despite this however, the two hypotheses, can, and indeed were applied to the path that DGIV needed to follow in order to pursue its particular goals. In ground-handling, the key question to be addressed is why it was the case that after so vehemently stating that it wished to use an Article 90 Directive, DGIV subsequently 'washed its hands' of the whole episode and allowed the Commission Proposal to become watered down in the way that it was (the Commission Proposal eventually led to a Council Directive which included the 'anti liberal' elements encapsulated in the Commission's Proposal).[88]

DGIV believed, and continues to believe, that ground-handling practices, as exercised by Frankfurt *et al* (where monopolistic ground-handling is alleged to exist), violate Article 86 of the Treaty of Rome. Furthermore (and crucially), DGIV believes that the ground-handling Directive does not stipulate sufficient legislation to comply with Article 86. However, following the report of the Comité des Sages on air transport liberalisation in January 1994 (a Committee set up by DGVII),[89] and, in particular, the ACI February 1994 Document, 'Consultations on Ground-handling Services',[90] it became clear that, because of political pressure, Van Miert would be unable to persuade a majority of his fellow Commissioners to formulate an Article 90 Directive. In particular, 'Germany said that it would not be desperately impressed if the Commission used an Article 90 Directive'.[91] Furthermore, Kassim adds, 'Ever since the near rejection of the Maastricht Treaty in France, the Commission has been very, very careful of its political responsibilities.'[92] Moreover, the Chief Executive of ADV stated, 'I had a long, one to one conversation with Van Miert [about this matter.]'[93] DGIV subsequently decided that it would be fruitless to follow the 'logical alternative', the path of a liberally-minded Council Directive on this issue, as Van Miert was also fully aware that a liberally worded Proposal would not achieve a Qualified Majority Vote at the Transport Council. An indication of DGIV's 'detachment' from this process is given by the fact that DGIV did not even oppose the fact that the eventual Commission Proposal was for a Council Directive rather than a Council Regulation.[94] When questioned as to why the Commission formulated a Proposal for a Directive, rather than a Regulation, an air transport lawyer stated:

> Because a Regulation is so much more effective. A Council Regulation is immediately applicable in all Member States, without the need for further legislation...but that would not be very pleasant for those states that did not want to liberalise. Such states wished for the delay in legislation to be as long as possible. That is why the only thing the Commission could achieve was a Directive, not a Regulation.[95]

ACI seems to have succeeded in two major areas prior to the Commission formulating any Proposal. The first was that ACI was successful in ensuring that the Proposal was ultimately for a Council, rather than a Commission Directive. Second, ACI ensured that the final legislation would take the form of a Council Directive, rather than the more efficient Council Regulation. The Head of the European Policy Unit at ACI stated, 'We were pleased that the policy was a Council Directive, rather than a Council Regulation.'[96]

DGIV's only remaining option was to use legal precedent in order to achieve a shift from a regulated to a liberalised ground-handling market. As the competition rules of the Treaty of Rome are directly effective, DGIV decided to focus on the legal nature of the original complaints in order to achieve its goals.

> The fact that we were strongly deterred from using an Article 90 Directive on ground-handling does not prevent us from using relevant case law - such as the current case against Frankfurt.[97]

The complaints from the airlines concerning Frankfurt were based on the allegation that Frankfurt was abusing its 'dominant market position'. The 'narrowly specified market' to which the airlines referred was the air-side ground-handling market at that airport. For the Commission or Court of Justice to find in favour of the complainants they must look at existing case law, initially to establish dominance, and subsequently to establish whether or not an abuse of that dominance had taken place.[98]

In such a situation, the key test case which establishes dominance is referred to - *Hoffmann-La Roche* (1979).[99] Furthermore, the 'demand substitutionality' (i.e. the choice that a consumer of ground-handling services has), is calculated based on relevant cases such as *Ahmed Saeed* (1989),[100] and *British Midland v. Aer Lingus* (1992).[101] Once dominance in the relevant market place had been established (Frankfurt Airport is a monopoly supplier of ground-handling services), the Commission or Court must find that abuse of the dominant market position has taken place.

The type of abuse which the complainant airlines alleged was 'refusal to supply access'. In this case the claim referred to the refusal to supply the provisions for self-handling or third party handling at this particular airport. Therefore, relevant cases (particularly those relating to port access, such as *Holyhead I* (1992),[102] were referred to. As stated, the Commission was under substantial (and successful), political pressure not to formulate a Commission Article 90 Directive. However, this did not prevent the Commission from formulating a Commission Decision (on Frankfurt) based on Article 86 of the Treaty. A Commission Decision is a key test of law in a particular circumstance.

A Commission Decision can be described as an 'invitation' to the subject of the complaint to pursue the particular matter in the Court of Justice. A Commission Directive is seen to be the Commission 'making policy'; thereby attracting the accusation of being a political action. Conversely, a Commission Decision is 'seen' as an interpretation of existing policy, open to challenge in the ECJ; use of such *jurisprudence* is not seen by Member States to be the Commission interfering in matters which are 'none of its concern', but rather an application of Community law. As stated in Chapter 2, there is in reality no great deal of difference between a Commission Directive and a Commission Decision, as both are interpretations of existing competition law. However, the former is often seen by both Member States and their composite interests (albeit inaccurately) as policy-making, whilst the latter is seen as policy implementation.

In January 1998 the European Commission found that, in the specific case of Frankfurt Airport, the undertaking had indeed abused its dominant market position and subsequently accepted the complaints of the airlines that they were prevented from providing self-handling, and that other third party handlers were prevented from offering their services. The

Commission rejected Frankfurt Airport's argument that 'constraints on space and capacity' justified the monopoly. Frankfurt Airport was given four months in which to lodge an appeal with the European Court of Justice.[103] Despite giving indications that it would launch such an appeal, Frankfurt Airport eventually conceded that such action would prove fruitless.[104]

What is most relevant about the *Frankfurt* Decision is that it goes beyond the stipulations laid out by the ground-handling Directive;[105] i.e. the Decision effectively stated that the Directive does not fully meet the competition criteria laid out by Article 86 of the Treaty. Therefore, DGIV have used *jurisprudence* in order to achieve its liberalising goals. It has also allowed DGIV to exercise the powers to implement its policies on ground-handling that were stated by Le-Fret *et al.* Primarily, it allows DGIV to achieve its goals via legal precedent, in order to bypass reluctant Member States. Furthermore, in October 1997 (prior to the *Frankfurt* Decision), the Commission ended an investigation into Olympic Airways arising from its monopoly activities at Athens Airport which also seemed to violate the competition provisions of the Treaty of Rome.[106] The Commission ended the investigation as Olympic, aware that the Commission would find against it and impose stiff financial penalties, amended many of its ground-handling procedures at Athens.[107]

DGIV subsequently created two situations on the issue of ground-handling; *Frankfurt,* which created a jurisprudence based on monopoly airport ground-handling activities, and Athens (although, in this case it was not the formulation of a Commission Decision *per se*, rather the discontinuation of an existing investigation), aimed at the monopoly ground-handling activities of the flag carrier airline.

The lobbying by the elite air transport interests was successful in preventing a Directive which would have ensured thorough competition in ground-handling. But their initial victory - the prevention of an Article 90 Directive being formulated by the Commission - may eventually be nullified by the *jurisprudence* created by the *Frankfurt* case. Therefore the lobbying by the elite airport interests was successful in initially preventing a Commission Directive and then in successfully influencing the formulation of the Commission Proposal for a Council Directive, which minimised the negative consequences of liberalisation. However, the lobbying by elites was ultimately not effective in preventing DGIV from pursuing its mandate.

Slot allocation

On this issue DGIV did not follow the same path in the attempt to achieve its goals. An analyst of the situation must therefore ask why DGIV does not employ similar tactics in the case of slot allocation? Why does DGIV not allow the Proposal to proceed in the way that DGVII wishes it to proceed and then apply *jurisprudence*, upon the policy becoming a Regulation if DGIV sees grandfather rights and secondary trading as an abuse of a dominant market position?

In the case of ground-handling, legal opinion widely agreed that the ground-handling Directive did not meet the competition requirements of the Treaty of Rome; this fact is even covertly accepted by many of the interests which so vehemently opposed ground-handling liberalisation. However, grandfather rights and secondary trading do not present such a clear legal case. It may well be argued that the large proportion of grandfather rights held by British Airways at Heathrow, Lufthansa at Frankfurt, et cetri, amounts to a dominance of the market position. However, whether or not this constitutes an abuse of a dominant market position is highly debatable. Grandfather rights appear not to amount to predatory pricing or unfairly high pricing, which are two of the legal tests of abuse detailed in Chapter 2. It may be argued in some quarters that grandfather rights may amount to a 'refusal to supply access to an essential facility'. Relevant cases in this area include *Holyhead II* (1994),[108] where the owners of the port of Holyhead prevented access to a competitor who wished to operate from its port. However grandfather rights are *de facto* owned by the airline (for example, British Airways), rather than the owners of the airport (for example, Heathrow's owners BAA), so this case is unlikely to be so relevant.

Subsequently, although DGIV and DGVII both agree that secondary trading is illegal under the auspices of the existing Regulation, it is unlikely that a Regulation allowing the activity would be open to challenge in the European Court of Justice. In a secondary trading environment, airlines would not be forced unwillingly to sell their slots to dominant carriers, and therefore secondary trading can hardly be argued to be an abuse of a dominant market position. Subsequently, although DGIV sees both grandfather rights and secondary trading as inherently anti-competitive practices, it is constrained by the fact that, unlike in the case of ground-handling, it is far less certain that it can achieve its desired outcomes in these areas by way of jurisprudence.[109]

Conversely, Competition Commissioner Van Miert has stated that he would 'not be surprised' to see cases at the ECJ on access to congested airports by new entrants'.[110] Furthermore, an official from DGIV stated that grandfather rights have the potential to be classed as a violation of the Treaty of Rome. However, this optimism is heavily qualified. The same official stated that violations would have to be proved, 'under very particular circumstances', adding,

> ..we must wait for the right case to arrive before taking any action... grandfather rights are a very difficult area to prove illegal under the auspices of EC competition law.[111]

Whatever the precise legal arguments, it is clear that DGIV is substantially less confident of its ability to overturn grandfather rights and secondary trading, than it is to overturn the ground-handling situation, should the issues go before the European Court of Justice. Therefore, rather than using *jurisprudence* DGIV needed to approach a different avenue to achieve its objectives.

In June 1996 British Airways and American Airlines announced that they were to form an air transport alliance. The proposed alliance was to be, what Philippe Chappate of Slaughter and May describes as a 'strategic alliance'.[112] Air transport alliances can take on many forms. Primarily, alliances are a result of national interference in the running of airlines. Ownership restrictions exist to block ownership by foreign nationals, and the only way to take advantage of the economies of scale which come about in international merger situations is to form an alliance. According to Philippe Chappatte, a strategic alliance would usually involve:

- Joint fare planning and budgeting.
- Route and schedule co-ordination (in particular, planning centred on each airline's hub airport).
- Code-sharing.[113]
- Reciprocal frequent flyer arrangements.
- Joint marketing, advertising, distribution and sales

A non-strategic alliance would be more limited and normally only include limited code-sharing. In my opinion it might also include joint marketing to a certain extent. But it would certainly not include joint fare planning, and reciprocal FFPs; the key issues of competition concern by DGIV of the proposed BA/AA alliance. The proposed BA/AA alliance indeed comprised the criteria of a strategic alliance.[114]

This announcement by British Airways and American Airlines added a new dimension to the disputed issues of secondary trading and grandfather rights, as well as to the relationship between the transport legislative and the *quasi-judicial* arms of the Commission. DGVII and DGIV viewed, and continue to view the proposed alliance as a benchmark test of their opposing views on the issues of secondary trading, and grandfather rights.

DGIV believes that the alliance would increase the dominance of British Airways and American Airlines at Heathrow, as the number of grandfather rights that the two airlines held there would be combined. Furthermore, the airlines would jointly set fares and according to DGIV this would make the alliance a *de facto* merger, despite the fact that there is to be no exchange of equity between the two carriers.[115]

DGIV demanded a 'surrender' of a proportion of the grandfather rights as a condition of the alliance (see below). Subsequently, British Airways demanded that if any slots were to be surrendered as a 'price' of the alliance BA must have the ability to be able to sell these slots on the open market.[116] DGIV has however repeatedly stated that it sees any such trade as potentially anti-competitive and has demanded that the slots must be surrendered without payment.

The Commission had previously formulated a Proposal for a Council Regulation in 1989 on such alliances.[117] Had the Proposal become a Regulation, it would have awarded the Commission the clear mandate to vet alliances between EU carriers and third countries (such as the US).

However, DGIV was convinced that Article 89 of the EC Treaty already gave the Commission the competence to vet the proposed alliance. Subsequently, following the announcement of the proposed alliance, the Commission declared that it too would be investigating the alliance, utilising for the first time Article 89.[118] The Commission also stated that it would be launching identical enquires into the other EU / transatlantic alliances already in place as well as the proposed alliance between BA and AA.[119] The retention of grandfather rights, and the ability of airlines to engage in secondary trading became, because of the resulting enquiries into the alliances, an increased matter of concern to many of the EU Association of European Airlines (AEA), members, themselves already in alliances with other US carriers. These alliances are:

- Lufthansa of Germany and United Airlines[120]
- KLM of the Netherlands and Northwest Airlines[121]
- Austrian Airlines / Swissair / Sabena (Belgium), and Delta Airlines[122]
- Scandinavian Airlines and United Airlines[123]
- British Airways / US Air, (investigation into which was discontinued upon the break-up of this alliance). [124]

Therefore DGIV is using the investigations into current and existing EU - US airline alliances in order to extrapolate as many grandfather rights, without financial compensation (i.e. without the possibility of secondary trading the grandfather rights), as the 'price' of these alliances. Such alliances have traditionally been dealt with by the regulatory authorities of the two states involved, for example, the UK and USA for the proposed British Airways / American Airlines alliance, Netherlands and USA, for the KLM / Northwest alliance, et cetri.

Over a year after the announcement of DGIV's investigation into the alliances, DGIV issued a 'Comfort Letter' to BA. In it Van Miert stated that BA and AA must:

- Surrender 353 slots at Heathrow without financial compensation (i.e. without the ability to secondary trade those slots)
- Cut capacity by 50% between Heathrow and Chicago, Miami, Boston and Philadelphia

• Freeze capacity for five years on flights to Baltimore, Seattle, and Charlotte
• Operate independently (i.e. no fare setting), on the London - Dallas route (a route where BA and AA have a duopoly).[125]

As a *quasi-judicial* unit, DGIV could make such demands on the alliance as it had the confidence that if its authority to do so was to be tested it would subsequently be recognised by the European Court of Justice. Indeed, this fact was not tested in the Court of Justice, presumably as a result of the certainty of a victory by DGIV and the acceptance of DGIV's competence in this area by associated interests.

Therefore, although DGIV is unconvinced that a challenge to grandfather rights or secondary trading would succeeded in the Court *per se,* it is certain that the Commission has the legal ability to demand the surrender of grandfather rights as the price of an alliance. However, this is quite different to the way in which the Commission may impose conditions in a merger situation.[126] In a merger situation, divisions of a company must often be sold off; but they are sold off for a price; they are not 'given' away in the way in which DGIV demands that BA 'gives-up' its slots at Heathrow. Subsequently, this is a unique demand for DGIV to make.

What Case Study No 2 shows is the way in which DGIV can go even beyond the jurisprudence, and 'secret jurisprudence', specified by Wilks in Chapter 2, in order to achieve its goals. The proposed alliance was an opportunity for DGIV to extend its competence in the area of EU-transatlantic alliances, using powers it already possessed under Article 89 of the Treaty of Rome. However, such 'pushing of the boundaries' of Commission competence inevitably led to a 'turf war' between the two sections of the Commission vying for competence in these particular areas. In short, the proposed alliance both magnified and publicised the divisions between DGIV and DGVII on slot allocation. The slot allocation episode demonstrates the extreme flexibility of DGIV, to 'stretch' the elastic of the competition provisions of the EC Treaty in order to maximise its competence in the face of seemingly irremovable political obstacles.

DGIV faced criticism initially from the then UK Conservative administration, as well as British Airways, that it had not investigated previous transatlantic alliances, yet chose to investigate them only when

BA / AA announced their intentions. In response the Director of DGIV stated:

> The Council [had] not adopted the necessary regulations (in 1989) so the Commission has fallen back on the original treaty (the Treaty of Rome), which provided Article 89 for the enforcement of competition rules (those rules being Articles 85 and 86).
>
> It is not for want of trying to secure a better legal framework that this step, necessary for the proper implementation of competition policy, was taken.[127]

Furthermore, Van Miert stated that the Commission had (prior to the announcement of BA / AA in June 1996), been reluctant to open procedures on the basis of Article 89 of the Treaty, which, he argued, was basically a transitional, 'fail safe', provision, 'But given the multiplication of alliances we finally had no other choice', he rather unconvincingly argued.[128]

Additionally, Van Miert also argued that prior to BA / AA there were few alliances, and their scope of their co-operation was not as great, and hence the impact of competition was more limited. However, this is unlikely to be the case. The extent of alliances such as the Lufthansa / United alliance equal, and in some circumstances exceed, the proposed alliance between BA and AA. For example, figures from the Official Airline Guide, display that Lufthansa and United have 63% of the slots at Frankfurt. KLM and Northwest have 59% of slots at Schiphol. However, BA / AA would have only 42% of the slots at Heathrow.[129]

It seems clear that DGIV is using Article 89 in order to achieve, at least in part, its goal of reducing the number of grandfather rights held by flag carrier airlines, without financial compensation. Indeed, Soames appears to argue that BA / AA was always going to be the case that the Commission used Article 89. This was quite obvious, he states. Yet Soames appears to argue that BA, and in particular the UK government, failed to see what the Commission was attempting to do and what powers the Commission had. He seems to indicate that it was a complete fiasco on behalf of the UK government.[130]

Lacking the confidence to use *jurisprudence* to achieve its goals, DGIV seized upon the opportunity presented to it with the announcement of the British Airways and American Airlines proposed alliance. As a direct result, the Proposal for the revised slot-allocation Regulation has been delayed for over three years, as the two issues of grandfather rights and secondary trading are central to any Proposal that is formulated. Furthermore, this disagreement has resulted in 'Commission inertia' due to the complete inability of the two directorates to reach a consensus.

Summary

The fact that the two hypotheses cannot be applied equally to DGIV and DGVII displays a great deal about the nature of the two arms of the Commission. As a *quasi-judicial* institution, DGIV holds a level of independence from interests (whether elites or non-elites), when drawing up its position on an issue, that is not enjoyed by DGVII. Despite this however, elite interests had a major effect on the form of 'policy making' which DGIV needed to undertake in order for it to pursue its goals.

In the ground-handling situation it became clear to DGIV that it would be unable to attain a simple majority of the College of the Commissioners in order to formulate a Commission Directive on ground-handling.[131] Despite this setback, DGIV was able to utilise the channel of jurisprudence in order to pursue its goals. Similarly, when faced with substantial opposition from elite interests on slot allocation, DGIV was able to use seemingly extraneous factors (such as the conditions imposed on airline alliances), in order to achieve its goals. The two paths followed by DGIV in these two different scenarios display the extreme flexibility of DGIV in responding to various situations. In particular, these case studies display DGIV's ability to 'circumnavigate' elite interests in a way which has not been previously examined. The use of *jurisprudence* in the case of ground-handling reflects previous attempts by DGIV to use *jurisprudence* in order to assert its authority in areas where it clearly felt that it possessed such authority. Rather than have any role in the formulation of a Council Directive on ground-handling (stemming primarily from the fact that DGIV was well aware that any radical liberalising elements of a Proposal would be vetoed by reluctant Member States), DGIV 'washed its hands' of the situation as soon as it became clear that it would be unable to formulate a Commission Directive.[132]

Similarly, the case study of the slot-allocation Regulation displays DGIV's ability to circumnavigate political barriers in order to achieve its goals. In this situation it was able, for the first time, to vet transatlantic airline alliances, using Article 89 of the EC Treaty. This provided a 'back door mechanism' for DGIV to achieve its objectives.

In summary, the application of hypothesis number 1 depends upon whether one is focusing on the legislative, or *quasi-judicial* nature of the Commission. A neo-liberal intergovernmentalism exists within the policy making of the legislative, but not within the policy making of the *quasi-judicial* unit of the European Commission. Hypothesis number 2 also depends upon whether one is examining the legislative or *quasi-judicial.* Greenwood's hypothesis of the necessity of a domination of elite interests within a group is indeed reflected upon DGVII; however, it is largely irrelevant when reflected upon the position taken by DGIV. Political spillover has clearly occurred in these areas; this is evident from the substantial amount of involvement that the Commission has in these policies. Yet the dynamics of policy formulation, in regards to the legislative DGVII in these case studies of air transport have hardly altered. The elites dominate its position-making and the elites within the group must be accommodated in order to achieve effective interest representation at this level. There exists therefore, a *neo-liberal intergovernmentalism* within DGVII. However, the dynamics of policy formulation are substantially different in regards to the *quasi-judicial.* Furthermore, the *quasi-judicial* DGIV will exploit all available avenues to achieve its goals, as the case studies clearly demonstrate.

Notes

1 Interview: Jean Faussurier (Expert à la C.E.), DGIV, Brussels, 29.1.98. Indeed it might possibly be argued that an over-emphasis was placed upon Frankfurt by critics; 96% of Europe's airports provided a monopoly on the air side (see Chapter 5).

2 David Gardner, Ronald Van de Krol, 'Brussels to probe service monopolies at airports', *Financial Times,* 24.8.93, p.2.

3 Interview: Corinne Dussart-Lefret (Senior Administrator, Transport), DGIV, Brussels, 10.4.97.

4 Interview: Jean Faussurier (Expert à la C.E.), DGIV, Brussels, 29.1.98.

5 See below.

6 Jonathan Faull (Director DGIV) and Frederick Sorensen (Head of Air Transport Division, DGVII), evidence to House of Commons Transport Select Committee, 'The proposed alliance between British Airways and American Airlines', 15.7.96.

7 Scott Crosby (Partner, Kemmler, Rapp, Böhlke and Crosby), 'Recent developments in slot-allocation', paper presented at the European Air Law Association, Copenhagen, 8.11.96.

8 Ibid, see also: Dinos Kyrou, 'Landing slots and property principles', *Financial Times* (letters), 5.2.97, p.14.

9 A common term to describe a market, the legal validity of which is not clear.

10 Robert Webb QC (Head of Chambers, 5 Bell Yard, London), 'The legal implications of slot trading', paper presented at the Royal Aeronautical Society, 'Slots and access to airport facilities', conference, London, 21.1.97.

11 Hugh O'Donovan (Partner, Wilde Sapte, UK), 'Dealing and trading in slots', paper presented at the Royal Aeronautical Society, 'Slots and access to airport facilities', conference, London, 21.1.97. The section of Regulation 95/93 to which this legal ambiguity is directed is: Article 8.4, 'Council Regulation on common rules for the allocation of slots at Community airports', O.J. L 14/1, 12.1.93, p.4.

12 European Commission, Directorate General for Transport (DGVII), 'Discussion Document-Revision of Council Regulation 95/93', Brussels, 11.4.96.

13 Interview: Jean Faussurier (DGIV), Brussels, op.cit.

14 Article 14, O.J. No L 14/1, 22.1.93, op.cit, p.6.

15 At the time of his paper O'Donovan highlighted that slot trading, and the practice of grandfather rights had never been tested in a national court of a Member State, or in the European Court of Justice. This situation has altered since this time. See Preface for further information.

16 There exists substantial confusion on the matter of Commission Press Releases. A Commission Press Release does not necessarily reflect the views of the Commissioners. Indeed, Commission Press Releases are often issued in order that the various sections of the Commission might try and offset one another. In the case study of the ground-handling Directive, the Press Release was solely issued by DGIV, without consultation with other sections of the Commission. Interview: Helmut Kreis (former acting head of transport section, DGIV), Brussels (via telephone), 26.2.98.

17 European Commission (Press Release), IP/93/714, 5.8.93.

18 i.e. the airlines wishing to self handle.

19 i.e. third party handling.

20 European Commission (Press Release), IP/93/714, 5.8.93.

21 See Chapter 2.

22 Corinne Dussart-Lefret and Christine Federlin (Senior Administrators, DGIV), 'Ground-handling Services and EC Competition Rules - A Commission initiative to open up ground-handling markets', in *Air and Space Law,* Vol.XIX., Number 2 (1994), p.53.
There are three primary ECJ cases which formed the basis to this claim:
Case C-41/90, *Hofner and Elser v. Macroton ('Hofner')* (1991) ECR I-1979;
Case C-179/90, *Merci convenzionali porto di Genoa Spa v. Siderurgica Gabrilli ('Port of Genoa')* (1991) ECR I-5889;

Case 311/84, *Centre Belge d'Etudes de Marche Telemarketing (CBEM), v. Compagnie Luxembourgeoise de Telediffusion and Information publicite Benelux* (1985) ECR 3261.

23 Corinne Dussart-Lefret and Christine Federlin, op.cit (1994).

24 ACI Europe, 'Consultations on Ground-handling Services. Response by ACI Europe to the European Commission DGIV and DGVII' (Feb, 1994).

25 Refer to Chapter 5.

26 Lefret and Federlin, op.cit (1994), pp.57-58.
DGIV also argued against the view that liberalisation would require unfeasible investment costs by a new independent ground handler (an argument put forward by ACI).

27 This was to be preceded by a Commission Proposal, and based on Article 84 of the EEC Treaty.

28 Lefret and Federlin, op.cit (1994), pp.57-58.

29 Ibid, p.60.

30 See: Trevor Soames (Partner, Norton Rose), 'Ground-handling Liberalisation', paper presented at IBC conference on EU air transport liberalisation, Brussels, 6.2.97; Commission Directives on competition are primarily formulated by DGIV. See Chapter 2, pp.57-77.

31 Until the lobbying pressure by various interests 'persuaded' DGIV to abandon the notion of a Commission Directive (see p.245), the draft Commission Directive remained in the hands of DGIV. When it became clear that DGIV would be unable to see such a Directive through the College of Commissioners, DGVII technically took control of the drafting the Proposal for a Council Directive to proceed through the College. However, DGVII, as a legislative unit seem to have allowed its Proposal to be 'shaped' by the dominant interests. DGIV, as a *quasi-judicial* unit, was not prepared to allow this to happen (interview: Corinne Dussart-Lefret (Senior Administrator, Transport), DGIV, Brussels, 10.4.97).

32 For example, see *Financial Times,* 24.8.93, op.cit.

33 Points 15, 16, 18, Explanatory Memorandum, Commission of the European Communities, 'Proposal for a Council Directive on access to the ground-handling market at Community airports', COM (94) 590 final, 13.12.94, pp.5-6.

34 The Selection Procedure is where potential ground handlers tender a bid to provide their services. The airport operator, however, does not need to tender a bid, as it has the guaranteed ability to provide ground-handling services, according to the Proposal, if it so wishes.

35 See Chapter 5.

36 Article 2.1., COM (94) 590 final, op.cit, p.21.

37 Articles 9.1., 9.2., 9.6., COM (94) 590 final, op.cit, pp.24-25. See also: Explanatory Memorandum, Point 16., pp.5-6.

38 Article 8, p.24, ibid.

39 Although it would technically be illegal for an airport to deliberately put in place such barriers to trade, proving that the airport is acting with anti-competitive intent in such cases would be exceedingly difficult for a complainant to prove. See Soames, op.cit.

40 Article 1.3, COM (94) 590 final, op.cit, p.21.

41 See Chapter 4, f/n No.43.

42 Article 7.2, COM (94) 590 final, op.cit, p.23.
43 Article 5, p.22, ibid.
44 See Chapter 5.
45 Article 10.2.(d), p.26, ibid.
46 Trevor Soames, 'Ground-handling Liberalisation', op.cit, p.14.
47 Article 7, COM (94) 590 final, 13.12.94, op.cit, p.23.
48 Article 6, COM (94) 590 final, op.cit, p.22.
49 Article 6.2. defines the 'flag carrier as, 'any user who has carried more than 25% of the passengers recorded at the airport during the year preceding', ibid, p.23.
50 Article 10.2., COM (94) 590 final, op.cit, p.26.
51 The Council Directive on ground-handling subsequently included all of the 'anti liberalising' features of the Commission Proposal, stipulated above. The process of the formulation of this Directive from the original Commission Proposal, via the Co-operation procedure of the European Parliament, to final Council Directive, is outside the scope of this research, as this book concentrates primarily on Commission dynamics. However, the various points of the legislative procedure are detailed in chronological order, here: (i) COM (94) 590 final, op.cit; (ii) European Parliament, Committee on Transport and Tourism, 'Report on the Proposal for a Council Directive on access to the ground-handling market at Community airports', A4-0272/95, 31.10.95; (iii) EP Debate No 4-470/208, 16.11.95. pp.212-219. and amendments: O.J. C. 323/94, 4.12.95., pp.94-107; (iv) European Commission, 'Amended Proposal for a Council Directive on Access to the Ground-handling Market at Community Airports', COM(96) 75 final, 12.03.96; European Council, 'Common Position (EC No.21/96) with a view to adopting Council Directive 96/.../EC on access to the ground-handling market at Community airports', O.J. C 134/30, 6.5.96, pp.30-48; (v) Council, 'Common Position (EC No.21/96) with a view to adopting Council Directive 96/.../EC on access to the ground-handling market at Community airports', O.J. C 134/30, 6.5.96, pp.30-48; (vi) E.P. Session Document, 'Recommendation for second reading on the common position established by the Council with a view to the adoption of a Council Directive on access to the ground-handling market at Community airports (C4-0220/96 - 94/0325(SYN))', 4.7.96; (Second Reading): E.P. Session Document, 'Decision on the common position established by the Council with a view to the adoption of a Council Directive on access to the ground-handling market at Community airports(C4-0220/96 - 94/0325(SYN))', 16.7.96; (vii) European Commission: COM (96) 435 final, 'Re-examined Proposal for a Council Directive on the ground-handling market at Community airports' 16.8.96; Council Directive 96/67/EC, of 15.10.96, on access to the ground-handling market at Community airports, O.J. L 272/36, pp.36-45; (viii) 'Council Directive on access to the ground-handling market at Community airports', O.J. L 272/36, 25.10.96.
52 Interview: Philip Lowe (Chef de Cabinet to Transport Commissioner Kinnock), Brussels, 17.4.97; Interview: Barry Seal MEP (*Rapporteur* on ground-handling legislation), European Parliament Transport and Tourism Committee, Brussels, 6.1.98; Peter Morrisore OBE (Managing Director, Airport Co-ordination Ltd.), Heathrow Airport, 10.1.97; Michael Niejhar (Senior Administrator, DGVII), 'Recent developments

in slot-allocation', paper presented at the European Air Law Association, Copenhagen, 8.11.96; Philip Lowe, in *Flight International* (2-8 Oct. 1996), p.5.

53 Interview: Corinne Dussart-Lefret (DGIV), op.cit, Brussels; Karel Van Miert (Competition Commissioner), 'Competition policy in the air transport sector', speech at the Royal Aeronautical Society, London, 9.3.98; Interview: Barry Seal (MEP), Brussels, op.cit; Peter Morrisore OBE (Airport Co-ordination Ltd.), op.cit, Heathrow; See also press, November 1996 onwards. For example: 'The long arm and big nose of Mr Van Miert', *The Independent* (Business Comment), 14.1.97, p.18; 'Van Miert denies air row is settles', *Financial Times*, 26.2.97, p.2.

54 A 'new entrant' airline is defined by Regulation 95/93 as an air carrier holding less than 3% of the slots at a particular airport. Additionally, other conditions are attached; see Article 2.b, O.J. L 14/1, 22.1.93, p.2. See also Chapter 4.

55 Coopers and Lybrand, 'The application and possible modification of Council Regulation 95/93 on common rules for the allocation of slots at community airports', 17.10.95, p.17.

56 Chris Castles (author of report, 'The application...', partner, Coopers and Lybrand), 'Development of airport slot-allocation regulation in the European Community', paper presented at the IBC, 'European Air Transport Services and Airports', conference, Brussels, 7.2.97.

57 Coopers and Lybrand, op.cit, p.16.

58 European Commission (DGVII), 'Discussion Document...', op.cit, p.1.

59 Coopers and Lybrand, op.cit, p.2.

60 Michael Niejhar (DGVII), 'Recent developments...', op.cit.

61 Interview: Senior source within DGIV.

62 'Council Regulation (EEC) No...amending Council Regulation No 95/93 of 18 January 1993 on common rules for the allocation of slots at Community airports (Explanatory Memorandum), Point 4, p.2.

63 Ibid, Point 5, p.2.

64 Neil Kinnock (Transport Commissioner) (letter), *Financial Times*, 7.4.97, p.22.

65 Philip Lowe (Chef de Cabinet to Kinnock), 'The new environment for EU air transport services', paper presented at IBC, 'European air transport services and airports' conference, Brussels, 6.2.97.

66 Interview: Philip Lowe, Brussels, op.cit.

67 *Financial Times*, 10.12.96, p.8.

68 Although he was stated as a 'source' in *The Independent*, it is quite possible (although by no means definite), that this source was Philip Lowe.

69 *The Independent*, 10.12.96, p.17.

70 Karel Van Miert (Competition Commissioner), 'Competition policy in the air transport sector', speech at the Royal Aeronautical Society, London, 9.3.98.

71 Interview: Barry Seal MEP, Brussels, op.cit.

72 Interview: Brian Simpson MEP (Member, Transport and Tourism Committee), Brussels, 20.1.97.

73 O.J. L 272/36, op.cit, 25.10.96.

74 See: Chapter 4.

75 See: Chapter 5.

76 Interview: Bervd Nierobisch (Executive Director), ADV, Stuttgart, 10.7.97.
77 ACI: 'The Economic Benefits of Air Transport' 1992, p.11.
78 Interview: Kees Veenstra (Deputy Secretary General), AEA, Brussels, 16.4.97.
79 See Chapter 5.
80 Greenwood, Chapter 1.
81 McLaughlin, Jordan and Maloney, Chapter 1, p.35.
82 See Chapter 5.
83 Jordan, McLaughlin, Chapter 1, pp.30-32.
84 Chapter 5.
85 See Olson's theory and response by Jordan and McLaughlin, Chapter 1, pp.31-33.
86 See Chapter 1.
87 See Chapter 1.
88 Council Directive 96/67/EC, of 15.10.96, on access to the ground-handling market at Community airports, O.J. L 272/36, pp.33-42.
89 See Chapter 5.
90 See Chapter 5.
91 Interview: Trevor Soames (Norton Rose), Brussels, 6.2.97.
92 Interview: Dr Hussein Kassim, Bareback College, University of London, London, 28.1.98.
93 Interview: Bervd Nierobisch (Executive Director), ADV, Stuttgart, 10.7.97.
The fact that the representative of the German airports was permitted a personal conversation with the Commissioner is highly unusual. The head of the European Policy Unit at ACI stated that the Chef de Cabinet and other officials always attend meetings with Commissioners, and that the agenda is fixed in advance; any change from the fixed agenda leads to an immediate abandonment of the meeting. A meeting on such a personal level is therefore highly unusual (interview: Shenna Gibson (Head of European Policy Unit), ACI Europe, Brussels, 10.4.97).
94 Interview: Jean Faussurier (Expert à la C.E.), DGIV, Brussels, 29.1.98. A Directive requires national legislation to be formulated in the Member States. There is therefore a time delay in comparison to a Regulation, which is binding from the time of a Council decision.
95 Interview: Trevor Soames (Norton Rose), Brussels, 6.2.97.
96 Interview: Shenna Gibson (Head of European Policy Unit), ACI Europe, Brussels, 10.4.97.
97 Interview: Jean Faussurier (Expert à la C.E.) DGIV, Brussels, 29.1.98.
98 See Chapter 2, for legal details.
99 *Hoffmann-La Roche*, ECJ 13.2.79, ECR 461, 520.
100 Case 66/86 *Ahmed Saeed* [1989], ECR 803 [1990] *Common Market Law Review*, Vol.4, No.102.
101 *British Midland v. Aer Lingus* [1992] O.J. L 96/34 [1993] *Common Market Law Review*, Vol.4. No.516.
102 *B&I Line plc v. Sealink Harbours Limited and Sealing Stena Limited:* [1992], *Common Market Law Review*, No.255.

103 See: European Commission (DGIV), Press Release, IP/98/27, 14.01.98; 'Frankfurt Airport', O.J. L 72, 11.03.98. Interview: Scott Crosby (air transport solicitor, representing Frankfurt Airport) (via telephone), Brussels, 9.2.98.

104 See: European Commission, Press Release, IP/98/794, 8.9.98, 'Frankfurt Airport changes policy on long-term ground-handling contracts'.

105 Nota Bene: The Directive included the 'anti liberalising elements' of the Commission Proposal.

106 European Commission: Press Release, 'The Commission obtains a significant improvement in the supply of ground-handling services at Athens airport: the Commission ends an infringement procedure begun in 1994', IP/97/876, 15.10.97.

107 Interview: Jean Faussurier (Expert à la C.E.), DGIV, Brussels, 29.1.98.

108 *Sea Containers v. Stena Sealink - Interim Measures:* O.J. L 15/8 [1994]

109 Interview: Jean Faussurier (Expert à la C.E.), DGIV, Brussels, 29.1.98.

110 Karel Van Miert (Competition Commissioner), 'Competition policy in the air transport sector', speech at the Royal Aeronautical Society, London, 9.3.98.

111 Interview: Jean Faussurier (Expert à la C.E.), DGIV, Brussels, 29.1.98.

112 See: Philippe Chappatte (Partner, Slaughter and May), 'The Alliance Between SAS and Lufthansa - A Model For Future Co-operation Between Airlines in Europe', paper presented at the European Air Law Association, Copenhagen, 8.11.96.

113 Code sharing has been defined as:

> ...an agreement between air carriers, whereby one party operates a particular service, while allowing another party to hold out at the same time the same service under that other party's designator code, in addition to or n place of the code of the operating carrier.

This is not the clearest of definitions! It essentially means the ability of an airline to sell seats on a 'partner' airline, using the same flight prefix, for example, Virgin (VS), selling seats on Continental (CO). This definition was devised for DGVII by the consultancy firm Stratagem, and can be found in: *Code sharing: A study into the consequences for the internal European Air Transport Market,* and cited in, Trevor Soames, 'Opportunities and Challenges in the Aviation's New World Order: The application of EU Competition Rules to Air Routes from the EU to Third Countries', paper presented at the European Air Law Association, Vienna, November 1998.

114 Probably the best comprehensive overview of the proposed alliance (albeit from the view of BA), can be found in the British Airways evidence to the House of Commons Transport Select Committee, 10.7.96.

115 See Chapter 2, for details of the Commission's competence on mergers.

116 British Airways and American Airlines submitted detailed evidence to the House of Commons Transport Select Committee supporting their proposal. See: British Airways / American Airlines evidence, 10.7.96.

117 European Commission: 'Proposal for a Council Regulation (EEC) amending Regulation (EEC) No 3975/87 of 14 December 1987 laying down the procedure for the

application of the rules on competition to undertakings in the air transport sector', O.J. No C 248/7, 15.9.89.

118 Article 88 has 'traditionally' been viewed as awarding competence in such areas to the national competition authorities of Member States. However, Article 88 is superseded by Article 89 in its effectiveness, which awards competence to the Commission when the Member States have not taken adequate action , subject to the competition provisions of the Treaty of Rome. For a discussion of the legal competence of the Commission in this area, see: John Balfour (Partner, Frere, Cholmeley, Bischoff, London), 'EC / US airline alliances - who regulates?', paper presented at IBC, 'European air transport and air services', conference, Brussels, 6.2.97. See also Chapter 2, pp.66-67.

119 BA and AA have now formed an alliance in the 'non disputed' areas, named 'One World' (other airlines include: Quantas, Finnair, Canadian, Iberia and Japan Airlines). For example, they co-operate in marketing. However, they are unable to fare set, and offer FFPs on trans-Atlantic routes, until a final decision has been reached by the Commission and the US Dept. of Transport. This will depend very much on the negotiations currently taking palace between the UK and US government on the revisions of the current bi-lateral agreement. Access to Heathrow is a particular sticking point in the negotiations.

120 Now called the Star Alliance. Other members include: SAS, Air Canada, Thai Airways and Varig.

121 Other members include: Alitalia, Continental (US), Braathens, and Kenya Airways. Air France have links with both the *KLM et al* alliance and the Austrian *et al* alliance. A chart of existing alliances can be found in, *Airline Business* (June 1998), p.45.

122 Now known as the 'Qualiflyer' group.

123 Now part of the Star Alliance.

124 European Commission (DGs IV and VII), evidence to House of Commons Transport Select Committee, 15.7.96, op.cit. Note: As stated in Chapter 1, p.44, Alitalia is currently in an alliance, with a US carrier (Continental). However, this is not a 'fare setting' alliance, and therefore the Commission have not launched an investigation into the situation. See also: Michael Skapinker, 'Brussels explains reasons for airline enquiry', *Financial Times,* 16.7.96, p.7.

125 These conditions were sent to BA by 'Comfort letter'. Additionally, these details were leaked to the press. See: 'EU terms threaten to ground BA alliance', *The Observer* (Business Section), 27.7.97, p.1; 'Brussels sets airline hurdle high', *Financial Times,* 28.7.97. As far as the author of this book is aware, DGIV had not decided upon the conditions for the above alliances (other than BA/AA), at the time of this comfort letter.

126 See Chapter 2.

127 Jonathan Faull (Director, DGIV), 'Commission frustrated in attempt to investigate airline alliances' (letter), *Financial Times,* 11.7.96, p.22.

128 Karel Van Miert (Competition Commissioner), 'Competition policy...', op.cit.

129 Figures for KLM / Northwest and Lufthansa / United, taken from the Official Airline Guide, 1995. Figures for BA / AA from BA's own figures. See: House of Commons Transport Select Committee, 'The Proposed Alliance Between British Airways and American Airlines, 30.7.96, p.xiii.

130 Interview: Trevor Soames (Partner Norton Rose), London, 2.4.98.

131 Interview: Senior sources within DGIV.
132 Ibid.

Conclusion

What have the previous chapters revealed about the hypotheses and the case studies? This Conclusion will firstly revise the main points of the book before progressing to examine this question.

Air Transport has traditionally been a state-centric matter. It was, (and arguably still is), an issue which states around the globe saw as a subject which should remain in the confines of bilateral agreements; the Member States of the European Community were no exception to this rule, yet this global practice or 'global cartel' as described by Pillai (1969)[1] was in direct confrontation with the competition provisions of the Treaty of Rome. It took, however, almost thirty years from the signing of the Treaty before the European Court of Justice, by way of *Nouvelles Frontières,* stipulated that these global practices were indeed violations of EC competition law. Moreover, the European Commission was instrumental in the 'behind the scenes' activity in persuading the French travel agent to pursue the matter through the European Court of Justice. This activity stemmed from the actions of DGs IV and VII; the former to pursue competition policy, the latter to expand its own competence in the area. In this way the DGs acted as 'midwives' of integration, in a form of cultivated spillover. Conversely, in the examined case studies of the formulation of the ground-handling Proposal, and the attempted formulation of a Proposal for a Regulation on slot-allocation, this consensus between the legislative and the *quasi-judicial* broke down. But why was this was the case? This is reviewed in more detail below.

As a reminder, liberal intergovernmentalism focuses on the prominence of elite, domestic interests in determining the positions of their governments at the inter-state level. An 'elite accommodation', as described by Hix, occurs as these interests are the most important domestic interests, and ordinarily the largest firms within the particular states. Furthermore, such firms are usually the 'national champions'; they are the firms which are, or have been, in public / state ownership and are the firms to which the greatest political importance is attached. From the mid 1980s many of these firms in the European Community began to move from public to private ownership (particularly in the UK and the Netherlands), yet their 'national champion status' and hence their 'eliteness' has remained. Furthermore, the propensity of states to discriminate in favour of these interests, even in the 'more liberal' markets of the EU has also remained, despite claims by governments to the contrary. Conversely, according to Sandholtz, the importance of the 'national champion' has been in decline, resulting from the necessity for large projects, such as the manufacture of civil aircraft by Airbus, to require co-operation between several Member States. The individual state can no longer 'go it alone' on such projects, goes the argument. However, Sandholtz does not dispute the fact that the national champion aerospace firms, although becoming less important *vis á vis* the actions of their individual states, retain their importance at the EU level, albeit, a combined importance between the firms of the particular consortium, rather than an individual importance. Therefore, in the example of Airbus, rather than a situation where the composite parts influence the actions of their individual Member States (i.e. British Aerospace determining the position of the UK government, Aérospatiale, the French government, Daimler Benz, the German government and CASA the Spanish government), there exists a situation where the eliteness of the firms is being 'transferred' to pan European policy actors, such as the European Commission.

Hypothesis number 1 examined in this book was a test of an application of domestic dominance of policy making in the area of air transport, reflected onto the Commission's Proposal making process. Furthermore, hypothesis number 2 examined whether such an 'elite accommodation' must also exist *within* groups in order for a group's representation to be successful; without which, according to Greenwood, such groups are 'doomed' to failure (Greenwood, 1994, pp.195-196).[2] Furthermore, hypothesis number 2 stated that the decision making within groups must be based on a decision making procedure pivoted on a form of majority voting; without this, according to Jordan and McLauglin, 'Groups...become paralysed because they cannot maintain consensus amongst conflicting interests' (Jordan and McLaughlin, 1991, p.49).[3]

However, contemporary interest representation literature has concentrated almost exclusively on 'traditional' notions of policy formulation. This can be summarised as the classic statement – 'the Commission proposes, and the Council disposes'. The literature[4] correctly points to the Single European Act and the Treaty on European Union as key points in time when those that wished to influence policy formulation substantially increased their representation in Brussels. The arrival of a 'simplified' Co-decision procedure for air transport policy upon the ratification of the Amsterdam Treaty (as has now happened), will also fundamentally alter the way in which air transport interests must approach policy formulation in order to be successful. The substantial significance of the European Parliament in this area upon ratification cannot be overstated. Indeed, the legal complexities on the issue of slot-allocation, if the Commission continues in its delay of the Proposal, and the potential for substantial further delay was detailed extensively in Chapter 2.[5]

As has been maintained within this book, the application of the hypotheses upon Commission actions may well depend upon whether one is examining the *legislative* or the *quasi-judicial* natures of the European Commission. DGIV is unique as a *quasi-judicial* institution due to its primary role as an enforcer of the existing competition legislation contained within the Treaty of Rome. This enormous competence stems from the unique abilities awarded to it by Council Regulation 17 of 1962.[6] Officially, DGIV requires a simple majority of the College of Commissioners in order to execute its authority. However, as Wilks points out, the use of jurisprudence built up since 1962, in particular over the past two decades, has given the legal go-ahead for DGIV to execute a wide range of responsibilities over a substantial arena of situations. Furthermore, the use of 'secret jurisprudence' in the vast majority of cases (an unofficial form of jurisprudence which, by its nature does not need College approval) has led to a situation where, according to Wilks, DGIV may be exercising a level of authority which was not envisaged by the six founder Member States upon Council ratification of Regulation 17 (Wilks, 1992, pp.30-33).[7]

The hypotheses tested in this book examined the demands of an elite group, the Association of European Airlines, a non-elite group, the European Regional Airlines Association, and a non exclusionary group, Airports Council International.

In the case study of the formulation of the Proposal for a Council Directive on ground-handling, the AEA produced a lowest common denominator position, resulting from the reluctance of its members to endorse a firm position on the issue. Because of the *de facto* decision making procedure of this group, the AEA produced the weakest position from the three groups examined, and, as a result, these composite interests were *competitive* in the phraseology of Huelshoff (1994, pp.255-279).[8] However ERA produced a firm position on the matter demanding the maximum amount of liberalisation, whilst ACI took a diametrically opposite view, demanding the minimum amount of liberalisation, and displayed total opposition to the possibility of a Commission Directive on the issue.

The case study of the attempted formulation of a Commission Proposal on slot-allocation was the study of a Proposal delayed for over two and a half years (at the time of writing), contravening EC law as stipulated by Article 14 of Regulation 95/93.[9] On this issue, the AEA openly supported grandfather rights, and, the evidence suggests, covertly supported the legalisation of secondary trading. (The AEA was fearful that this was, in any case, under investigation, as the practice, exercised by all EU flag carriers, was contrary to Regulation 95/93, according to DGIV and DGVII). ERA shifted position on the matter of secondary trading, in order, it seems, to appease the dominant interests, the 'subsidiary carriers' within its own grouping, whilst ACI was largely a 'non-player' on the issue.

In the former case study the views of ACI were reflected by the legislative DGVII in its Proposal, but not by the *quasi-judicial* DGIV, which more closely reflected the views of the non-elite group, the ERA. In the latter case study, DGVII favoured the views of the AEA. The only interest whose view was reflected by DGIV on slot-allocation was the airline Virgin Atlantic.

Application of hypotheses

On ground-handling, the thesis held that a form of liberal intergovernmentalism, a neo-liberal intergovernmentalism, was indeed apparent in the position making of DGVII, but not DGIV. What does such a finding reveal to the analyst?

First, such a finding reflects the fact that eliteness is a relative concept. Clearly, the views of DGVII more closely reflected the views of ACI than the AEA. Although the substantial eliteness of the flag carrier airlines was examined in Chapter 4, when compared directly to the airports in the Member States, the number of personnel directly employed by the airport (a key test of eliteness, according to the criteria used within this book), is, approximately six times greater than that of the flag carrier airline.[10] Furthermore, the airports achieved support from their Member States which the airlines, in many cases, did not necessarily receive. Therefore, when faced with a 'head on' challenge between two interests which are, in themselves, elite, the 'level of eliteness' becomes a measured concept by the policy formulator. Eliteness is relative to the opposition of other interests in a particular situation; it is not an absolute concept.

In regard to the application of hypothesis number 2, ACI felt that it must continue to accommodate the views of the German airports group ADV; not to have done so may have led to a withdrawal of ADV from ACI. ACI felt that such action by ADV would substantially weaken its interest representation to policy formulators. ADV clearly was, and continues to be, an elite in a group which itself consists of elite and non-elite interests. The Greenwood hypothesis was employed inasmuch as ACI realised that not to accommodate the views of such a significant interest would directly result in its own representation being severely undermined.

Furthermore, ADV applied hypothesis number 2 within itself. ADV conducted decision making with a form of qualified majority voting. When the decision on ground-handling had been made within ADV, all of its member airports undertook not to 'break from the pack' and not to separately organise their own interest representation. ADV subsequently became the most effective sub-group within the ACI structure. Moreover, ADV utilised its substantial leverage over policy formulators in order to pursue policy, not only at Level II (within the group), but also within Level 1 (the state, in this case, the German state), and Level III (independently to EU policy formulators). Additionally ADV's actions at Level III representation reflected the form of co-operative interest representation described by Huelshoff; groups in a co-operative bargaining position have a substantial advantage in promoting their views. ADV is a clear example of this.

Both the AEA and ACI formed lowest common denominator positions, yet the latter was substantially more successful in its interest representation than the former. Why was this the situation? Within the AEA, composite interests, in particular, the airlines Olympic, Iberia and Tap, 'broke from the pack' of interest representation. Despite the fact that the AEA formed a position that all its members 'could live with' some airlines circumnavigated the AEA in order to represent themselves to policy formulators. However, ACI's Level III representation was complementary to the lowest common denominator position, which was established as a result of pressure from ADV. In the case of the AEA, the position of its members Olympic, Iberia and Tap, via their Level III interest representation, was contradictory to its lowest common denominator position. Subsequently, one may ascertain that whether or not a group produces a position by a lowest common denominator or by a majority is largely irrelevant; what is crucial is that firms do not 'break from the pack', thereby upholding the 'paralysis' hypothesis of Jordan and McLaughlin (1991).[11] Furthermore, such findings at least partly address how interest groups can avoid the dilemma put forward by UNICE in Chapter 1,

> European legislators are likely to receive so many diverging messages from so many different interest groups that they could end up either very confused or able to divide and rule...[12]

The 'strength' of the position of ACI, resulting at least in part from the consolidation of its views on ground-handling did not, however, make a difference regarding the position of DGIV. As can be seen from the quotes and documentation cited in Chapter 6,[13] DGIV remained resolutely in favour of ground-handling liberalisation, and diametrically opposed to the views of ACI. Conversely, the position of ACI fundamentally affected the way in which DGIV could go about pursuing its policy. It became apparent to DGIV in 1994 that, because of political pressure, it would not be able to achieve its goal of a Commission Directive on ground-handling, as it would be unable to achieve a simple majority of the College of Commissioners. This would appear to be a victory for the neo-liberal intergovernmentalist in relation to both the legislative and the *quasi-judicial* arms of the Commission on this issue. DGVII was successfully lobbied, as can be seen from its highly 'illiberal' Commission Proposal for a Council Directive. DGIV was prevented from launching a Commission Directive, which would have dealt with the issue in a thorough manner. However, as Chapter 6 detailed, such a perception of the success of neo-liberal intergovernmentalism would prove to be superficial. The case study of the ground-handling Directive built upon Wilks's argument, promoted in Chapter 2, of the way in which DGIV has massively expanded its own competence, arguably beyond what Regulation 17 of 1962 originally intended. Indeed the case study supported Wilks's hypothesis that DGIV is able to use jurisprudence in order to pursue its goals. In this case, aware that it could not achieve a simple majority at the College, and aware that it could not hope to influence a Proposal for a Council Directive (a Proposal which would have had to have passed through a Council which, on the whole, was hostile to its aims), DGIV 'washed its hands' of this approach, and decided to support the original complaint made against Frankfurt and other airports by British Airways and other airlines. According to Jean Faussurier of DGIV, DGIV hoped to use the jurisprudence which came from this ruling in order to execute legislation which, in itself, would override the Council Directive on ground-handling. The Commission passed a Decision on this matter on 14 January 1998.[14] However, a question remains; why was it the case that the College of Commissioners felt unable to pass a Commission Directive on ground-handling (although a Directive was never actually formally voted upon), yet felt able to pass a Decision against Frankfurt?

As Chapter 2 of this book stipulated, both a Commission Decision or a Commission Directive on ground-handling would be applications of existing EC competition law. A Commission Directive on ground-handling would have applied to all the airports in the EEA; the Commission Decision addressed the anti-competitive behaviour of a particular airport - Frankfurt. However, the Decision against Frankfurt creates a jurisprudence which the Commission can follow in all other similar cases referred to it. Subsequently, the Decision will affect the cases against Rome, Milan, Madrid and other airports about which complaints have been made as well as potential violators. Subsequently, the net effects of a Commission Directive or a Commission Decision on ground-handling are very similar. However the rationale behind the College's thinking on this issue (and subsequently an attempt to answer the above question), is analysed later in this Conclusion.

In the case of ground-handling, *neo-liberal intergovernmentalism* affected the position of DGVII, a legislative element of the European Commission. The group cohesion of ACI, and its ability to accommodate its most prominent member, ADV, was crucial in determining the effectiveness of its interest representation to DGVII. Although the position of ACI was indeed one formed by a lowest common denominator, it was a position where its composite interests were content not to 'break from the pack'. However, *neo-liberal intergovernmentalism* cannot be applied to the position of the *quasi judicial* arm of the European Commission, DGIV, as the views of the elite interests were not reflected by this Directorate. However (and crucially), both hypotheses 1 and 2 played a crucial role in determining the path that DGIV needed to follow in order to achieve its goal of a liberalised ground-handling market. Liberal intergovernmentalism and group cohesion fundamentally altered the way in which DGIV approached this issue; but ultimately, by use of jurisprudence, DGIV was able to circumnavigate the demands of the elite interests, and the effectiveness of the group cohesion of ACI.

The path followed by DGIV is substantially different in the case study of slot-allocation. Before examining why this was let us review some of the aspects of this case.

The non-elite group, the ERA, switched its position on secondary trading. The appointment of an air transport policy manager has resulted in a situation where the subsidiaries of the elite flag carriers, although in themselves a numerical minority, seem to have been appeased within this interest group. This is, therefore, an application of the Greenwood hypothesis (domination of elites within the group), within a group which is, in itself, non-elite. This 'elite accommodation' means that the position of ERA largely reflects the position taken by the AEA. However, as admitted by ERA there was no actual vote to establish this position. The position appears to reflect the views of its 'subsidiary airlines' of the flag carriers. Therefore, the question must be asked: why did such an action not lead to a group paralysis? Why do the airlines that do not wish to adopt this particular policy lobby at Level III (independently to the Commission), in order to pursue their particular goals? The answer to this question lies within Chapter 1 of this book. Non-elite interests do not possess the resources, either in finance, manpower or 'know-how' to operate at Level III, or indeed at Level 1 (to their own Member State). Non-elites are therefore Level II (interest group) reliant. As this is the case, it could be argued that the Greenwood thesis is somewhat easier to operate in a group which is, itself, non-elite. Within such groups there exists less of a possibility that the member firms will 'break from the pack', as they simply do not have the resources to take such action.

In the slot-allocation case study, a neo-liberal intergovernmentalism exists inasmuch as the view of DGVII reflects a near unanimity of interest representation; although it can hardly be argued that it reflects elite over non-elite, as the only interest which publicly opposes grandfather rights and secondary trading is the airline Virgin Atlantic. But what of the position of DGIV in relation to the application of the two hypotheses? Again, hypothesis number 2 proved to be largely irrelevant. Although ACI remained largely neutral on the issue, the views of the AEA and ERA were directly and publicly opposed by DGIV; therefore, group cohesion played no part in the formation of the view of DGIV.

Furthermore, the position of DGIV on the matter can, in no way be argued to be an application of neo-liberal intergovernmentalism; therefore hypothesis number 1 is also rejected in relation to DGIV.

However, the path which DGIV was forced to follow did not reflect the path of jurisprudence which was followed in the ground-handling episode. This was because DGIV was unable to use jurisprudence, resulting from the uncertainty of the 'illegality' of secondary trading and grandfather rights. DGIV therefore had no option but to use another avenue, that of the forced surrender of grandfather rights, but without the ability to secondary-trade those rights, as the price of various alliances between EU carriers and US carriers. DGIV saw this as an expansion of the competence which it already had under Article 89 of the Treaty. It is likely that that the announcement of the proposed alliance between British Airways and American Airlines presented a 'window of opportunity' for DGIV to, at least in part, withdraw grandfather rights from 'dominant' airlines without financial compensation. However, when DGIV first stated its conditions to British Airways, by way of a 'comfort letter', British Airways stated that such conditions[15] would mean that British Airways would 'walk away' from the deal rather than have such conditions imposed.[16] There then followed a 'game' of negotiation between British Airways and DGIV. It was in the interests of DGIV that the alliance went ahead as it realised that this may be the only occasion that it could extract a 'price' (i.e. grandfather rights without financial compensation), from British Airways, and other carriers in alliances. However, too high a price would lead to a situation where BA would 'walk away' from the alliance (and other airlines might possibility 'walk away' from alliances already in place), if they were to calculate that the financial cost of such a surrender of slots outweighed the financial gains that they would achieve from their alliances. On 7 July 1998, the European Commission, for the first time in an official document stipulated the conditions by which the BA/AA alliance could proceed. It also stated the conditions by which the United / Lufthansa alliance could continue. British Airways were instructed that they must surrender 267 slots at Heathrow without financial compensation (plus other conditions),[17] although DGIV's competence in this area has been threatened by Lufthansa, which has been ordered to surrender 108 weekly slots mostly at Frankfurt.[18] Lufthansa has threatened to pursue DGIV to the European Court of Justice. However, the overwhelming bulk of legal opinion is that the Commission has the competence in this area to demand such surrender as a price of an alliance. DGIV continues to investigate KLM / Northwest, SAS / Lufthansa, and Delta / Swissair / Sabena / Austrian.

In conclusion to this book, the hypotheses as stipulated in the Introduction - an existence of a form of liberal intergovernmentalism, and the necessity of group cohesion for successful interest representation at the level of the European Commission, are highly dependent upon whether one is lobbying the legislative or the *quasi-judicial* arms of the European Commission. Officially, the actions of DGIV require the consent by a majority of the College of Commissioners. Although the College may not be as beholden to elite interests as the Member States, it is still influenced by them as was most clearly demonstrated in the ground-handling situation where a majority of Commissioners stated informally to Van Miert that they would not support him on a Commission Directive on ground-handling. However, DGIV was able to use jurisprudence in order to circumnavigate this politicisation of the Commission. In the case of slot-allocation, DGIV was able to use Article 89 of the Treaty, again circumnavigating elite interests, and stating that the extraction of grandfather rights was a 'price' of these alliances. This method of 'stretching the elasticity of competence' goes beyond the jurisprudence or even the 'secret jurisprudence' stated as so vital by Wilks for DGIV's success in the area of competition. DGIV has been able to gradually view the competition provisions of the Treaty of Rome in an increasingly wide context, thereby bypassing much of the traditional interest representation theory.

There remain, however, significant questions which were not answered by this book. Chief amongst these questions is why it was that, in the case of the slot-allocation Regulation, Competition Commissioner Van Miert was able to carry a majority of the College with him in his insistence that grandfather rights be surrendered as a 'price' of the alliances, yet Transport Commissioner Kinnock was equally convinced that he could carry the support of the College to propose a Regulation allowing the secondary trading of grandfather rights, and their full retention, so opposed by DGIV. Similarly, in the case study of the ground-handling Directive, Van Miert was able to achieve a majority in favour of a Decision against Frankfurt, yet could not attain a majority for a Commission Directive which would have also dealt directly with this problem. I do not claim to have full explanations to these questions. However, the author has, from interviews, ascertained the following.

Commissioners do not, on the whole, wish to 'meddle' in areas that they see as belonging to other Commissioners. For most issues it is clear where the mandate of a particular Commissioner, and the relevant DG lie. The competence of DGIV, however, cuts across many of the Directorates of the Commission.

The issue of the revision of the 1993 slot-allocation Regulation is 'seen' by the majority of the College as primarily a transport issue; subsequently, the majority of Commissioners will support the Transport Commissioner in his view. However, the issue of surrendering slots as the 'price' of an alliance is 'seen' by the majority of the College as a DGIV issue; subsequently, it appears that the majority of Commissioners supported Van Miert in his view on the matter. However, as has been ascertained throughout this book, the two issues are intrinsically linked. It appears that it is because of this linkage, and because of the determination of Commissioners not to tread into the 'paths' of other Commissioners, that a gridlock was created between Van Miert and Kinnock. Kinnock was aware that if he proceeded with his Proposal, Van Miert would extract a very high price from the proposed and existing EU-transatlantic alliances.

In the case of the ground-handling Directive, it seems that the issue was 'seen' as one which was more closely associated with DGVII than DGIV, as it was 'seen' to be a transport policy in the making. As a result, DGIV's views on the matter were somewhat bypassed at the Proposal stage, as we have seen. Yet, when the Commission formulated a Decision on Frankfurt, arising from the complaints by the aforementioned airlines, this issue was 'seen' as one which was more closely associated with Van Miert's department than Kinnock's; subsequently, the Commissioners supported the wishes of Van Miert.

Why would Commissioners act in such a manner? It may be argued that Commissioners are attempting to 'isolate' or 'safeguard' themselves from lobbying pressure on matters which are not of direct concern to them. By applying this 'unwritten rule' within the Commission (not to 'tread on the toes' of other Commissioners in 'their' policy areas), Commissioners may well become less vulnerable to interest representation outside their direct area of responsibility. The problem (or the opportunity), for Van Miert and DGIV is that competition law cuts across the path, at some time, of the other members of the College of Commissioners.

This work has merely introduced the policy-making dynamics within the Commission, and the relevance of position making by groups and their relevance on Commission actions. There is a substantial amount of research, which needs to be conducted in at least six major areas.

The first area is the issue of Commission behaviour. It became apparent to the author of this book, via interviews, that the outcome of both of the case studies were, and are, being influenced largely by individuals within DGs IV and VII, and within the Cabinets of the Commissioners who wish to achieve certain goals to establish their own positions within the Directorates.[19] There is really very little wrong with this. In any successful organisation, individuals will try to promote themselves, and it is to the credit, not the detriment, of Commission personnel that they are no exception to this rule. However, this 'human factor' of Commission behaviour has been significantly overlooked. Too often the literature focuses on processes and outcomes, without enough analysis of the individuals which produce those outcomes. This factor (which has, coincidentally, been evaluated for some years in the literature on organisational behaviour), needs to be further examined by EU analysts.

Second, the nature of bargaining that takes place within the College (especially informal bargaining, 'log rolling' or 'favour swapping' between Commissioners), is an area which requires substantial further research in order to establish a correlation between issue linkage and Commission Proposals.

Third, the slot-allocation issue was deliberately delayed from the time of the first Regulation in 1993. It has become apparent to the author that the Commission has, and continues to deliberately delay matters which it cannot agree on (either formally or informally) within the College. However, such issues eventually need to be addressed. Further research needs to ascertain what the 'cost' of such delay is to policy formulators, interests and other actors.

Fourth, this work has touched upon Huelshoff's theory of co-operative and competitive interest representation, in relation to the domestic governance within a particular state. This work examined the role of the German interest group ADV as a co-operative interest group, utilising its 'weak' domestic government (using the criteria of Huelshoff), thereby producing a 'strong position' for ADV, and subsequently for ACI. Further data in this area is needed and further research needs to be conducted in order to find further linkage between Huelshoff's theory and actual case studies.

Fifth, Baumgartner and Leech, in their assessment of US policy formulation, point to a fundamental problem of attempting to 'prove' that the interest group activity of X causes policy Y;

> ...much recent quantitative research on lobbying and influence has been distilled into a narrow, tractable issue...This question, while perhaps not trivial, is far more narrow than the underlying question about interest group influence.[20]

Essentially, Baumgartner and Leech state that there are fundamental difficulties in proving the link between action and outco me. However, further research in this area of 'action and outcome' in the European Union will help to create a larger, more accurate picture for those researching policy formulation. That is not to say that this 'picture' will ever be finally complete, but further research would certainly lead to a greater understanding.

Baumgartner and Leech also criticise past research on interest group lobbying behaviour, as the research, they argue, tends to fall into one of two camps. The first camp is the 'broad based survey'. This form of research tends to be over-generalised and lack specific data, they claim. The second camp is the policy case study which includes such a limited number of issues that generalisation of the interest group system is 'all but impossible'.[21] Baumgartner and Leech conclude that scholars must somehow solve this problem of how to build in sensitivity to the contexts of decision making (achieved in the second of their examples), whilst combining broad surveys of group behaviour (achieved in the first). Clearly, research of this kind is needed in order to improve the analytical framework of interest group behaviour and lobbying activity within the European Union. However, the author of this work would dispute the claim by Baumgartner and Leech that: 'Unless groups play important roles in politics and policy making, ...their internal workings are irrelevant - nothing more'.[22] The reasons why groups may not play important roles (as analysed by Greenwood *et al*), are as important as establishing why various groups do indeed play important roles in policy making. Analysis of the 'why not' is as crucial as examination of the 'why' if one is to gain a comprehensive and advantageous examination of interest representation.

Finally, this work has also only begun to address the nature of DGIV as a *quasi-judicial* unit and its response to interest representation. Very little work has been conducted in this area, largely because the actions of DGIV are seen as legalistic in nature; subsequently, political scientists remain cautious of addressing the subject for fear of treading into 'alien' territory. Similarly, legal interpretations and studies of DGIV do not adequately address the political nature of competition policy, and in particular, the avenues which DGIV must follow to circumnavigate political obstacles. It will be to the detriment of such academic study if the two disciplines do not formulate a joint approach for the further study of the *quasi-judicial* nature of DGIV.

The difference between the *quasi judicial* and legislative natures of the Commission were not contained in either hypothesis of this work; rather the difference has become apparent to the author as a result of the research into the hypotheses. As a result of the research this book has developed this distinction.

This book has addressed the relationship between Commission Proposal making and the form of interest representation applied to the Commission. It has ascertained that the *quasi-judicial* element of the Commission, DGIV is an under examined object of study. Furthermore and crucially, it has established that this *quasi-judicial* element of the Commission, although not separated into a political vacuum from elite interests, has the ability to stretch and expand its competence, based on the existing competition provisions of the Treaty of Rome; an ability which has been overlooked by contemporary interest representation literature; this field is ripe for further research.

Notes

[1] See, 'Introduction', and Chapter 4.
[2] See Chapter 1.
[3] See Chapter 1.
[4] Extensively cited in Chapter 2, f/n Nos. 1-8.
[5] See Chapter 2.
[6] See Chapter 2.
[7] Ibid.
[8] See Chapter 1.
[9] Article 14, O.J. L 14/1, 22.1.93, p.6.
[10] Calculation from Table 6.1, Chapter 6.
[11] See Chapter 1. (See also: McLaughlin, Jordan and Maloney, 1993, p.199.)

[12] UNICE bulletin, No.1, 1991; also see Chapter 1.
[13] See Chapter 6.
[14] See: 'Frankfurt Airport', O.J. L 72, 11.03.98.
[15] See Chapter 6.
[16] Chris Allen (Head of Competition and Industry Affairs, British Airways), 'The Role of Code Sharing and Franchising in British Airways' Development'; paper presented at Royal Aeronautical Society, 'Franchising, Code Sharing and Related Matters' conference, London, 11.2.98.
[17] See: Commission Press Release, IP/98/64, and 'Commission notice concerning the alliance between British Airways and American Airlines', O.J. C 239, 30.7.98; Commission Press Release, IP/98/640 and 'Commission notice concerning the alliance between Lufthansa, SAS and United Airlines', O.J. C 239, 30.7.98.
[18] See: Commission Press Release, IP/98/640 and 'Commission notice concerning the alliance between Lufthansa, SAS and United Airlines', O.J. C 239, 30.7.98.
[19] Interview: Source of significant authority.
[20] Frank R. Baumartner and Beth L. Leech, *Basic Interests – The Importance of Groups in Politics and in Political Science* (Princeton: Princeton University Press, 1998, p.146).
[21] Ibid, p.165.
[22] Ibid, pp.187-188.

Bibliography

Bernardine Adkins, *Air Transport and EC Competition Law* (London: Sweet and Maxwell, 1994).

Aerospace (monthly journal of the Royal Aeronautical Society, 4 Hamilton Place, London, WC2) (various editions).

Airline Business (Published by Reed Business Publications, Sutton, Surrey) (various editions).

David Allen, 'Competition Policy: Policing the Single Market', in H. Wallace and W. Wallace, *Policy-Making in the European Union* (Oxford: Oxford University Press, 1996), pp.157-183.

Svein Andersen and Kjell A. Eliassen, 'Complex Policy-Making: Lobbying the EC', in Svein S. Andersen and Kjell A. Eliassen (eds), *Making Policy in Europe* (London: Sage, 1993), p.35-54.

Clive Archer, *Organizing Europe - The Institutions of Integration* (2nd ed) (London: Edward Arnold, 1994).

W.F. Averyt, *Agropolitics in the European Community* (London, Praegar, 1977).

David A. Baldwin (eds), *Neorealism and Neoliberalism - The Contemporary Debate* (New York, Columbia University Press, 1993).

John Balfour, 'Air Transport - A Community Success Story?', in *Common Market Law Review*, Vol. 31 (1994), pp.1027.

John Balfour, *European Air Law* (London: Butterworths, 1994).

B.H. Baltagi, JM. Griffin and DP Rich, 'Airline Deregulation: The Cost Pieces of the Puzzle', in *International Economic Review*, Vol.36, No.1 (Feb, 1995), pp.245-260

Howard Banks, *The Rise and Fall of Freddie Laker* (London: Faber & Faber, 1982).

Sean D. Barrett, 'Aviation and the Pursuit of Contestability', in *Economic Affairs*, Vol. 14 (Feb, 1994), pp.11-18.

Frank R. Baumgartner and Beth L. Leech, Basic Interests - The Importance of Groups in Politics and in Political Science (Princeton: Princeton University Press, 1998).

Frank R. Baumgartner and Bryan D. Jones, *Agendas and Instability in American Politics* (Chicago: University of Chicago Press, 1993).

R. Beuter and P. Tsakaloyannis (eds), *Experiences in Regional Co-operation* (Maastricht: European Institute of Public Administration, 1987).

Simon Bishop, 'Antitrust Enforcement and the Rule of Law', in *European Competition Law Review*, Vol.19, No1 (Jan, 98). pp.1-3.

Simon Bishop, 'The Modernisation of DGIV', in *European Competition Law Review,* Vol.18, No.8 (Nov, 1997), pp.481-484.

Simon Bishop, 'The European Commission's Policy Towards State Aid - A Role for Rigorous Competitive Analysis', in *European Competition Law Review,* Vol.18, No.2 (Mar, 1997), pp.84-88.

Geert P.A. Braam, *Influence of Business Firms on the Government - An Investigation of the Distribution of Influence in Society* (The Hague: Mouton, 1981).

Richard Branson, 'Market Access in the Airline Industry', in *Economic Affairs,* Vol.14, No. 2 (Feb, 1994), pp.6-10.

L.J. Brinkhorst, L.J. and Henry G. Schermers, *Judicial remedies in the European Communities: a Case Book* (Deventer: Kluwer, 1977).

S. Budd and A. Jones, *The European Community, A Guide to the Maze* (London: Kogan Page, 1989).

Simon J. Bulmer, 'Air Transport liberalisation' (typescript), University of Manchester, 1997.

Simon J. Bulmer, 'The Governance of the European Union: A New Institutionalist Approach', in *Journal of Public Policy,* Vol.13, No.4 (1993), pp.351-380.

Simon Bulmer, *The Domestic Structure of EC Policy Making in West Germany* (New York: Garland, 1986).

JR Burton, CD Zick and R.N. Mayer, 'Consumer Views of the Need for Government Intervention in the Airline Market', in *Journal of Consumer Affairs,* Vol.27, No1 (Summer, 1993), pp.1-22.

A. Butt Philip, 'Pressure Groups in the European Community and Informal Institutional Arrangements', in R. Beuter and P. Tsakaloyannis (eds), *Experiences in Regional Co-operation* (Maastricht: European Institute of Public Administration, 1987), p.75.

Alan Butt Philip, 'Pressure Groups in the European Community' (paper presented at the University Association for Contemporary European Studies, London, 1985).

Kenneth Button and Dennis Swann, 'European Community Airlines - Deregulation and its Problems', in *Journal of Common Market Studies,* Vol. XXVII, No.4 (Jun, 1989), pp.259-282.

A. Cawson, *Organized Interests and the State* (London: Sage, 1985).

C. Chant, *Airlines of the World* (London: Tiger Books, 1997).

The Chartered Institute of Transport in the UK, 'Kinnock bids for mastery of Europe's skies', in *Global Transport,* Autumn, 1996, pp.80-81, and Autumn 1995, 'Battle for the Open Skies', pp.48-49.

P. Conradt, *The German Polity* (5th ed) (New York: Longman, 1993).

John C. Cooper, *The Right to Fly* (New York: Henry Holt, 1947).

Alison Corke, *British Airways: The Path to Profitability* (London: Pinter, 1986).

M. Crozier and E. Friedberg, *Actors and Systems* (Chicago: Chicago University Press, 1980).

R.M. Cyert and J.G. March, 'Organisation, Structure and Pricing Behaviour in an Oligopolistic Market', in *American Economic Review* (Mar, 1955).

A. Dashwood, 'The Principle of Direct Effect in European Community Law', in *Journal of Common Market Studies,* Vol.16 (1978), p.245.

Wolfgang Deselaers, 'Liberalisation of Ground Handling Services at Community Airports', in *Air and Space Law,* Vol.XXI, No.6 (1996), pp.260-266.

Desmond Dinan, *Ever Closer Union?* (London: Macmillan, 1994).

Alan P. Dobson, *Flying in the Face of Competition* (Aldershot: Abebury, 1995).

Christopher Docksey and Karen Williams, 'The European Commission and the execution of Community policy', in Geoffrey Edwards and David Spence (eds), *The European Commission* (2nd ed) (London: Cartermill, 1997), pp.125-153.

M. Donnelly, 'The Structure of the European Commission and the Policy Formation Process', in S. Mazey and J. Richardson (eds), *Lobbying in the European Community* (Oxford: Oxford University Press, 1993), p.79.

D. Druckman, 'Determinants of Compromising Behaviour in Negotiation - A Meta-Analysis', in *Journal of Conflict Resolution,* Vol.38, No.3 (Sept, 1994), pp.507-556.

Andrew Duff (eds), *The Treaty of Amsterdam* (London: Sweet and Maxwell, 1997).

Corinne Dussart-Lefret and Christine Federlin, 'Ground Handling Services and EC Competition Rules', in *Air and Space Law* (Vol. XIX), No.2 (1994), pp.50-61.

The Economist (various editions) (Economist Group, London and Brussels).

Geoffrey Edwards and David Spence, *The European Commission* (2nd ed) (London: Cartermill, 1997).

Jan Ernst C. de Groot, 'Code-sharing - United States' policies and the lessons for Europe', in *Air & Space Law',* Vol.XIX. No.2 (1994), pp.62-74.

Peter Esaiasson and Soren Holmberg, *Representation From Above* (Aldershot: Dartmouth, 1996).

European Voice (various editions) (Economist Group, London and Brussels).

Peter B Evens, Harold K. Jacobson and Robert D. Putnam, *Double-Edged Diplomacy - International Bargaining and Domestic Politics* (Berkeley: University of California Press, 1993).

Nicholas Fraser, Philip Jacobson, Mark Ottaway and Lews Chester, *Aristotle Onassis* (London: Book Club Associates, 1978).

Flight International (Various editions) (Reed Business Publications, Sutton, Surrey).

Stephen George, 'Policy Networks' (unpublished paper, presented at the University of Aberdeen, 31.5.97).

Stephen George, *Politics and Policy in the European Community* (2nd ed) (Oxford: Oxford University Press, 1992).

Eileen M. Gleimer, 'Slot Regulation at High Density Airports: How Did We Get Here and Where Are We Going?', in *Journal of Air Law and Commerce*, Vol. 61, No. 4 (May-June 1996), p.890.

K. Goldmann and G. Sjostedt (eds), *Power, Capabilities, Interdependence* (London: Sage, 1979).

Wyn Grant, *Business and Politics in Britain* (Basingstoke: Macmillan, 1993).

W. Grant, 'Pressure Groups and the European Community', in S.Mazey and J. Richardson (eds), *Lobbying in the European Community* (Oxford: Oxford University Press, 1993), pp.27-46.

Wyn Grant, 'DIY: The Government Relations Functions of Large Companies', in Grant Jordan (eds), *The Commercial Lobbyists* (Aberdeen: Aberdeen University Press, 1991), pp.99-110.

F.I. Greenstein and N.W. Polsby (eds), *Handbook of Political Science*, Vol.8 (Reading, Massachusetts: Addison-Wesley, 1975).

Justin Greenwood, 'The Regulation of Interest Representation', in Justin Greenwood, *Representing Interests in the European Union* (Basingstoke, Macmillan, 1997), pp.90-100.

Justin Greenwood (eds), *European Casebook on Business Alliances* (London: Prentice Hall, 1995).

Justin Greenwood and Andrew M. McLaughlin, 'The Management of Interest Representation in the European Union', in *Journal of Common Market Studies,* Vol.33, No.1 (Mar. 1995), pp.143-156.

J. Greenwood, 'Pharma and Biotech: Virtues and Trends in EU Lobbying', in R.H. Pedler and M.P.C.M. Van Schendelen (eds), *Lobbying the European Union* (Aldershot: Dartmouth Publishing, 1994), pp.183-198.

Justin Greenwood, Jurgen R. Grote and Karsten Ronit (eds), *Organized Interests and the European Community* (London: Sage, 1992).

Justin Greenwood, Jurgen R. Grote and Karsten Ronit, 'Introduction: the Transnational Dimension', in Justin Greenwood, Jurgen R. Grote and Karsten Ronit (eds), *Organized Interests and the European Community* (London: Sage, 1992).

Paolo Guerrieri and Pier Carlo Padoan (eds), *The Political Economy of European Integration - States, Markets and Institutions* (London: Harvester Wheatsheaf, 1989).

K.M. Gwilliam, 'EEC Transport Policy Issues' (paper resented at Rotterdam Transport Centre, Erasmus University, Rotterdam, 1992).

Ernst B. Haas, 'Why Collaborate? - Issue-Linkage and International Regimes', in *World Politics* (April, 1980), Vol:XXXII, No.3, pp.357-405.

Ernst B. Haas, 'Turbulent fields and the theory of regional integration', in *International Organization* (Spring, 1976), Vol.30, No2, pp.173-212.

E.B. Haas, 'The study of regional integration: reflections on the joy and anguish of pretheorizing', in L.N. Lindberg and S.A. Scheingold (eds), *Regional*

Integration: Theory and Research (Cambridge, Massachusetts: Harvard University Press, 1971), p.6.

Ernst B. Haas, 'International integration: the European and universal process', in *International Organization*, Vol.15 (1961), pp.366-92.

Fiona Hayes-Renshaw and Helen Wallace, *The Council of Ministers* (Basingstoke: Macmillan, 1997).

Jack Hayward (eds), *Industrial Enterprise and European Integration - From National to International Champions in Western Europe* (Oxford: Oxford University Press, 1995).

John P. Heinz, *The hollow core: private interests in national policy making* (Massachusetts: Harvard University Press, 1993).

K. Hendricks, M Piccione and G. Tan, 'The Economics of Hubs: The Case of Monopoly', in *Review of Economic Studies,* Vol.62 (1995), pp.83-99.

D. Hickie, 'Airbus Industry and the Struggle in the GATT', in R.H. Pedler and M.P.C.M. Van Schendelen (eds), *Lobbying the European Union* (Aldershot: Dartmouth Publishing, 1994), pp.41-65.

Paul Hitchings, 'Access to International Telecommunications Facilities', in *European Competition Law Review,* Vol.19, No.2 (Feb, 1998), pp.85-98.

Simon Hix, 'The Study of the European Community: The Challenge to Comparative Politics', in *West European Politics,* Vol.17, No.1 (Jan,1994), pp.1-30.

Claudia Hubschmid and Peter Moser, 'The Co-operation procedure in the EU: Why was the European Parliament influential in the decision on car emissions standards?', in *Journal of Common Market Studies,* Vol.335, No.2 (Jun, 1997), pp.225-242.

M.G. Huelshoff, 'Domestic Politics and Dynamic Issue Linkage: A Reformulation of Integration Theory', in *International Studies Quarterly,* Vol.38 (1994), pp.255-279.

Barry B. Hughes, 'Delivering the Goods - The EC and the Evolution of Complex Governance', in Ray, D. Smith and J.L. Ray (eds), *The 1992 Project and the Future Integration in Europe* (London: M.E. Sharpe, 1993), pp.45-69.

R. Hull, 'Lobbying Brussels, A View From Within', in S. Mazey and J. Richardson (eds), *Lobbying in the European Community* (Oxford: Oxford University Press, 1993), p.83.

Barry Humphreys, 'Developments in Transport Policy - The UK Civil Aviation Authority and European Air Services Liberalisation' (typescript) (Feb, 1996).

James Humphreys, *Negotiating in the European Union* (London: Century, 1997).

Francis Jacobs and Richard Corbett, *The European Parliament* (London: Longman Group, 1990).

Jane's / Collins, *Civil Aircraft* (Glasgow: Harper Collins, 1996).

Christer Jönsson, *International Aviation and the Politics of Regime Change* (London: Pinter, 1987).

Christer Jönsson, 'Sphere of Flying: The Politics of International Aviation', in *International Organization,* Vol.35, No.2 (Spring, 1981), pp.73-302.

Grant Jordan, 'Negotiated regulation in North Sea Oil Exploration: The Case Study of Brent Spar' (typescript) (June 1998).

Grant Jordan (eds) *The Commercial Lobbyists* (Aberdeen: Aberdeen University Press, 1991).

Grant Jordan and Andrew McLaughlin, 'The Rationality of Lobbying in Europe: Why are Euro Groups So Numerous And So Weak?' (Centre for European Studies, Discussion Paper No.7, November, 1991).

Grant Jordan, 'The Pluralism of Pluralism: An Anti-theory?', in *Political Studies,* No.XXXVIII (1990), pp.286-301.

Hussein Kassim and Anand Menon (eds), *The European Union and National Industrial Policy* (London: Routledge, 1996).

Hussein Kassim, 'Air Transport', in, Hussein Kassim and Anand Menon (eds), *The European Union and National Industrial Policy* (London: Routledge, 1996), pp.107-131.

Hussein Kassim, 'Air Transport Champions: Still Carrying the Flag', in Jack Hayward (eds), *Industrial Enterprise and European Integration - From National to International Champions in Western Europe* (Oxford: Oxford University Press, 1995), pp.189-214.

H. Kassim, 'Policy Networks, Networks, and European Union Policy Making: A Sceptical View', in *West European Politics,* Vol.17, No.4 (Oct, 1994), pp.15-27.

P. Katzenstein (eds), *Between Power and Plenty* (Madison: University of Wisconsin Press, 1978).

R.O. Keohane and S. Hoffmann (eds), *The New European Community* (Oxford: Westview Press, 1991).

Robert O. Keohane and Joseph S. Nye, 'International Interdependence and Integration', in F.I. Greenstein and N.W. Polsby (eds), *Handbook of Political Science*, Vol.8 (Reading, Massachusetts: Addison-Wesley, 1975), pp.363-409.

Christopher Kerse, 'Enforcing Community Competition Policy Under Articles 88 and 89 of the EC Treaty', in *European Competition Law Review',* Vol.18, No.1 (Feb, 1997), pp.17-23.

Neil Kinnock, 'The Liberalisation of the European aviation industry', in *European Business Journal* (1996), pp.8-13.

Emil J. Kirchner, 'The European Community: Seeds of Ambivalence', in Gordon Smith, William E.Paterson, Peter H.Merkl and Stephen Padgett (eds), *Developments in German Politics* (London: Macmillan, 1994), pp.172-184.

B. Kohler Koch, 'Patterns of Interest Intermediation in the European Union', in *Government and Opposition*, Vol.29, No.2 (Spring 1994), p.171.

S. Krasner, 'Approaches to the State: Alternative Conceptions and Historical Dynamics', in *Comparative Politics*, Vol.16 (1984), pp.223-224.

V.A. Kremenyuk (eds), *International Negotiation - Analysis, Approaches, Issues* (Oxford: Jossey-Bass, 1991).

Dinos Kyrou and James Thompson, *European Domination by Incumbents? Air Transport and Telecommunications – A Reflection of Commission Decision Making,* paper presented at the University Association for Contemporary European Studies, Research Conference, The University of Sheffield, September, 1999.

D. Kyrou, 'Air Transport Policy in the European Union - State Control By Other Means?', paper presented at the University Association for Contemporary European Studies, Research Conference, Loughborough University, September, 1997.

Brigid Laffan, *Integration and Co-operation in Europe* (London: Routledge, 1992).

T.D. Lairson and D. Skidmore, *International Political Economy* (London: Harcourt Brace College Publishers, 1993).

Laraine Laudati, 'The European Commission as regulator: the uncertain pursuit of the competitive market', in Giandomenico Majone, *Regulating Europe* (London, Routledge, 1996), pp.229-261.

Dick Leonard, *The Economist Guide to the European Community* (London: The Economist Books, 1992).

A. Lijphart, 'Democratic Political Systems: Types, Cases, Causes and Consequences', in *Journal of Theoretical Politics,* Vol.1, No.1 (1989), pp.499-515.

A. Lijphart, *Democracy in Plural Societies: A Comparative Exploration* (New Haven: Yale University Press, 1977).

A. Lijphart, 'Typologies of Democratic Systems', in *Comparative Political Studies* (1968), pp.3-44.

Leon N. Lindberg, 'Financial Deregulation, Monetary Policy Co-ordination and Economic and Monetary Union', in, Ray, D. Smith and J.L. Ray (eds), *The 1992 Project and the Future Integration in Europe* (London: M.E. Sharpe, 1993), p.117.

L.N. Lindberg and S.A. Scheingold (eds), *Regional Integration: Theory and Research* (Cambridge, Massachusetts: Harvard University Press, 1971).

Juliet Lodge (eds), *The European Community and the Challenge of the Future* (2nd ed) (London: Pinter, 1995).

Juliet Lodge, 'EC policymaking: institutional dynamics', in Juliet Lodge (eds), *The European Community and the Challenge of the Future* (2nd ed) (London: Pinter, 1995), pp.1-36,

Juliet Lodge, 'The Single European Act and the New Legislative Co-operation Procedure: A Critical Analysis', in *Journal of European Integration,* Vol.XI, No.1 (1987).

Andreas F. Lowenfeld, 'A New Takeoff for International Air Transport', in *Foreign Affairs,* No.1 (Oct, 1975), pp.37-50.

T.J. Lowi, *The End of Liberalism* (New York: W.W. Norton, 1979).

G. McConnell, *Private Power and American Democracy* (New York: Alfred A.Knopf, 1967).

Andrew M. McLaughlin, Grant Jordan and William A. Maloney, 'Corporate Lobbying in the European Community', in *Journal of Common Market Studies,* Vol.31, No.2 (Jun, 1993), pp.191-211.

Ian McLeod, *Judicial review* (Chichester: Rose Law, 1993).

Giandomenico Majone, *Regulating Europe* (London, Routledge, 1996).

William A. Maloney, Grant Jordan and Andrew M. McLaughlin, 'Interest Groups and Public Policy: The Insider / Outsider Model Revisited', in *Journal of Public Policy,* Vol.14, No.1 (1994), p.25.

P.L. Marín, 'Competition In European Aviation: Pricing Policy And Market Structure', in *The Journal of Industrial Economics,* Vol.XLIII, No.2 (June 1995), pp.141-159.

Peter Martin and Elizabeth Martin (Shawcross and Beaumont Solicitors) (eds) *Air Law*, Fourth Edition, Vol.2 (London: Butterworths, 1998).

Sonia Mazey and Jeremy Richardson, 'Pressure groups and lobbying in the EC', in, Juliet Lodge (eds), *The European Community and the Challenge of the Future* (2nd ed) (London: Pinter, 1995), p.37.

S. Mazey and J. Richardson (eds), *Lobbying in the European Community* (Oxford: Oxford University Press, 1993).

Peter H. Merkl and Stephen Padgett (eds), *Developments in German Politics* (London: Macmillan, 1994).

Neil J. Mitchell, *The Conspicuous Corporation - Business, Public Policy and Representative Democracy* (Michigan: University of Michigan Press, 1997).

J. Mo, 'The Logic of Two-Level Games with Endogenous Domestic Coalitions', in *Journal of Conflict Resolution*, Vol.38, No.3 (Sept, 1994), pp.402-422.

Andrew Moravcsik, 'Liberal Intergovernmentalism and Integration: A Rejoinder', in *Journal of Common Market Studies,* Vol.33, No.4 (Dec, 1995), pp.611-628.

Andrew Moravcsik, 'Preferences and Power in the European Community: A Liberal Intergovernmentalist Approach', in *Journal of Common Market Studies*, Vol.31, No.4 (Dec, 1993), pp.473-524.

Andrew Moravcsik, 'Negotiating the Single European Act: National Interests and Conventional Statecraft in the European Community', in *International Organization,* Vol.45 (Winter, 1991), pp.19-56.

J.L. Mortensen, 'The Interaction Between European Politics and International Bargaining: The Case of the Uruguay Round', paper presented at the European University Institute, Florence (Dec, 1995).

Brent F. Nelsen and Alexander C.G. Stubb (eds), *The European Union - Readings on the Theory and Practice of European Integration* (London: Lynne Rienner, 1994).

Michael Nentwich and Gerda Falkner, 'The Treaty of Amsterdam: Towards a New Institutional Balance', in *European Integration online Papers (EIoP),* Vol.1, No.15. www.eiop.or.at/eiop/

Michael Newman, *Democracy, Sovereignty and the European Union* (London: Hurst and Co, 1996).

Sir William Nicoll, 'The Code of Conduct of the Commission Towards the European Parliament', in *Journal of Common Market Studies,* Vol.34, No2 (Jun, 1996), pp.275-282.

Sir William Nicoll and Trevor C. Salmon, *Understanding the New European Community* (Hemel Hempstead: Harvester Wheatsheaf, 1994).

Michael Niejahr and Giuseppe Abbamonte, 'Liberalization policy and state aid in the air transport sector', in *Competition Policy Newsletter,* Vol.2, No.2 (Summer 1996).

E. Nordlinger, *On the Autonomy of the Democratic State* (Massachusetts: Harvard University Press, 1981).

Neill Nugent, *The Government and Politics of the European Union* (3rd edition) (Basingstoke, Macmillan, 1994).

Joseph S.Nye (eds), *International Regionalism* (Boston: Little, Brown, 1968).

M. Olson, *The Logic of Collective Action* (Mass: Harvard University Press, 1965).

Dolores O'Reilly, 'Testing Integration Theories: The Development of EU Air Transport Policy' (paper presented at the University Association for Contemporary European Studies, 11.9.97).

Dolores O'Reilly, 'From State-Centric Control to EC Competence', paper presented at European University Institute, Florence, 19.3.97.

W.E. Paterson, 'Regulatory Change and Environmental Protection in the British and German Chemical Industries', in *European Journal of Political Science,* Vol.19 (1991), pp.307-26.

J.P. Peckstadt, S. Mazey and J. Richardson, 'Defending and Promoting a Sectoral Interest within the European Community: The Case of the Textile Polyolefins Industry', in S. Mazey and J. Richardson (eds), *Lobbying in the European Community* (Oxford: Oxford University Press, 1993), pp.177-187.

R.H. Pedler and M.P.C.M. Van Schendelen (eds), *Lobbying the European Union* (Aldershot: Dartmouth Publishing, 1994).

J. Peterson, 'Policy Networks and European Union Policy Making: A Reply to Kassim', in *West European Politics,* Vol.18, No.2 (Apr. 1995), pp.389-407.

M.A. Petetaf and R Reed, 'Pricing and Performance in Monopoly Airline Markets', in *Journal of Law and Economics,* Vol. XXXVII (April, 1994), pp.193-213.

K.G.J. Pillai, *The Air Net - The Case Against the World Aviation Cartel* (New York: Grossman, 1969).

J. Pinder, 'Positive Integration and Negative Integration: Some Problems of Economic Union in the EEC', in *The World Today,* No.4 (1968), pp.88-110.

Mark A. Pollack, Creeping Competence: The Expanding Agenda of the European Community', in *Journal of Public Policy,* Vol.14, No.2 (1994), pp.95-145.

D. Puchala, 'Of Blind Men, Elephants and International Integration', in *Journal of Common Market Studies,* Vol.10 (1972), pp.267-285.

Robert D. Putnam, 'Diplomacy and domestic politics: the logic of two level games', in *International Organization*, Vol.42, No.3 (Summer 1988), pp.427-461.

Emmette S. Redford, 'A Case Analysis of Congressional Activity: Civil Aviation, 1957-58', in *Journal of Politics*, Vol.22 (1960), pp.228-258.

Peter Reed, 'Flying into trouble', in *New Economy,* Vol.2, No.1 (Spring, 1995), pp.9-13.

Philip A. Reynolds and Robert D. McKinlay, 'The Concept of Interdependence: Its Uses and Misuses', in K. Goldmann and G. Sjostedt (eds), *Power, Capabilities, Interdependence* (London: Sage, 1979), pp.141-169.

A.J. Reynolds-Feighan, 'European and American Approaches to Air Transport Liberalization - Some Implications for Small Communities', in *Transport Research, Policy and Practice,* Vol.29, No.6 (1995), pp.467-483.

Barry J. Rodger and Stuart R. Wylie, 'Taking the Community Interest Line: Decentralisation and Subsidiarity in Competition Law', in *European Competition Law Review,* Vol.18, No.8 (Nov, 1997), pp.485-491.

R. Rogowski, *Commerce and Coalitions - How Trade Affects Domestic Political Alliances* (Princeton: Princeton University Press, 1989).

James N. Rosenau, *The Scientific Study of Foreign Policy* (London: Frances Pinter, 1980).

W. Sandholtz, *High Tech Europe - The Politics of International Co-operation* (Berkeley, University of California Press, 1992).

A.M. Sbragia, 'Asymmetrical Integration in the European Community - The Single European Act and Institutional Development', in D. Smith and J.L. Ray (eds), *The 1992 Project and the Future Integration in Europe* (London: M.E. Sharpe, 1993), pp.92-109.

P.C. Schmitter and W. Streeck, 'Organised Interests and the Europe of 1992', in A.C.G. Stubb (eds), *The European Union - Readings on the Theory and Practice of European Integration* (London: Lynne Reinner, 1994), pp.169-187.

Gerrit Schohe, 'Rivals in a state', in *Airline Business,* Vol. 12, No. 8 (Aug, 1996), pp.42-45

J. Shaw, *European Community Law* (London: Macmillan, 1993).

H.A. Simon, 'A Behavioural Theory of Rational Choice', in *Quarterly Journal of Economics* (Feb, 1958).

P.J. Slot and P.D. Dagtoglou (eds), *Toward a Community Air Transport Policy - The Legal Dimension* (Boston: Kluwer, 1989).

Gordon Smith, *Democracy in Western Germany - Parties and Politics in the Federal Republic* (3rd ed) (New York: Holmes and Meier, 1986).

Ray, D. Smith and J.L. Ray (eds), *The 1992 Project and the Future Integration in Europe* (London: M.E. Sharpe, 1993).

David Spence, 'Structure, functions and procedures in the Commission', in Geoffrey Edwards and David Spence, *The European Commission* (2nd ed) (London: Cartermill, 1997), pp.103-124.

Martin Staniland, 'The Vanishing National Airline?' (typescript), Spring, 1998.

David Starkie, 'Developments in Transport Policy- The US Market in Airport Slots', in *Journal of Transport Economics And Policy,* Vol XXVIII, No.3 (Sep,1994), 325-329.

Handley Stevens, 'Liberalisation Of Air Transport In Europe', London School of Economics, European Institute paper (June 1997).

Jenny Stewart, 'Corporatism, Pluralism and Political Learning: A Systems Approach', in *Journal of Public Policy,* Vol.12, No.3 (Jul-Sep, 1992), pp.243-255.

Ian Stockman, 'Ramping up the price', in *Airline Business,* Vol.12 (Nov, 1996), pp.53-58.

John M. Stoppford and Susan Strange, *Rival States, Rival Firms - Competition for World Market Shares* (Cambridge: Cambridge University Press, 1991).

Susan Strange, 'Who are EU? Ambiguities in the Concept of Competitivness', in *Journal of Common Market Studies,* Vol.36, No.1 (March 1998), pp.101-114.

Susan Strange, *The Retreat of the State* (Cambridge: Cambridge University Press, 1996).

W. Streeck and P.C. Schmitter, 'From National Corporatism to Transnational Pluralism: Organised Interest in the Single European Market', in *Politics and Society,* Vol.19, No.2 (1991), pp.133-164.

A.C.G. Stubb (eds), *The European Union - Readings on the Theory and Practice of European Integration* (London: Lynne Reinner, 1994),

Secondo Tarditi, Kenneth J.Thomson, Pierpaolo Pierani and Elisabetta Croci-Angelini (eds), *Agricultural Trade Liberalization and the European Community* (Oxford: Clarendon Press, 1989).

Jeppe Tranholm-Mikkelsen, 'Neo-functionalism: Obstinate or Obsolete? A Reappraisal in the Light of the New Dynamism of the EC', in *Millennium: Journal of International Studies,* Vol.20, No.1 (1991), pp.1-22.

John Turner (eds), *Businessmen and Politics - Studies of business activity in British politics, 1900-1945* (London: Heinemann, 1984).

I. Van Bael, 'The Antitrust Settlement Practice of the European Commission', in *Common Market Law Review* (1986), No.23, pp.69-90.

Rene Van den Polder, 'Lobbying for the European Airline Industry', in R.H. Pedler and M.P.C.M. Van Schendelen (eds), *Lobbying the European Union* (Aldershot: Dartmouth Publishing, 1994), pp.103-119.

M.P.C.M. Van Schendelen, 'The Council Decides': Does the Council Decide?', in *Journal of Common Market Studies,* Vol.34, No4 (Dec, 1996), pp.531-548.

M.P.C.M. Van Schendelen, 'Introduction: Studying EU Public Affairs Cases: Does it Matter?', in R.H. Pedler and M.P.C.M. Van Schendelen (eds), *Lobbying the European Union* (Aldershot: Dartmouth Publishing, 1994). p.12.

David Vogel, *Kindred Strangers - The Uneasy Relationship between Politics and Business in America* (New Jersey: Princeton University Press, 1996).

Karen Walker, 'Bespoke fortunes', in *Airline Business,* Vol. 13. No. 1 (Jan, 1997), pp.32-35.

Helen Wallace and William Wallace, *Policy-Making in the European Union* (Oxford: Oxford University Press, 1996).

Helen Wallace, William Wallace and Carole Webb (eds), *Policy Making in the European Communities* (Chichester: John Wiley and Sons, 1977).

Carole Webb, 'Introduction: Variations on a Theoretical Theme', in Helen Wallace, William Wallace, Carole Webb (eds), *Policy Making in the European Communities* (Chichester: John Wiley and Sons, 1977), pp.1-31.

Martin Westlake, 'The Commission and the Parliament', in Geoffrey Edwards and David Spence, *The European Commission* (2nd ed) (London: Cartermill, 1997), pp.239-263.

Martin Westlake, *The Council of the European Union* (London: Catermill, 1995).

Martin Westlake, *A Modern Guide to the European Parliament* (London: Pinter, 1994).

Martin Westlake, *The Parliament and the Commission: Partners and Rivals in the European Policy-Making Process* (London: Butterworths, 1994).

Greg Williams, *The Airline Industry and the Impact of Deregulation* (Aldershot: Abebury, 1994).

P.J. Williamson, *Corporatism in Perspective: An Introductory Guide to Corporatist Theory* (London: Sage, 1989).

Daniel Wincott, 'Institutional Interaction and European Integration: Towards an Everyday Critique of Liberal Intergovernmentalism', in *Journal of Common Market Studies,* Vol.33, No.4 (Dec, 1995), pp.597-609.

Donncadh Woods (DGIV), 'An Outline of Community Competition Policy' (paper presented to Irish Institute for European Affairs, University College Dublin, 22.3.96),

(on, http://europa.eu.int/en/comm/dg04/speech/six/en/sp96008.htm)

Peyton H.Young, *Negotiating Analysis* (Michigan: University of Michigan Press, 1991).

EEC/EC/EU documentation, (reverse chronological order)

European Commission Press Release, 'Frankfurt Airport changes policy on long-term ground handling contracts', IP/98/794, 8.9.98.
'Commission notice concerning the alliance between British Airways and American Airlines', Official Journal of the European Communities, C 239/10, 30.7.98.
'Commission notice concerning the alliance between Lufthansa, SAS and United Airlines', O.J. C 239/5, 30.7.98.
European Commission Press Release: 'Commission publishes its conditions for approving the British Airways / American Airlines air alliance', IP/98/641, 7.8.98.
European Commission Press Release: 'Commission publishes its conditions for approving the Lufthansa/SAS/United Airlines air alliance', IP/98/640, 7.8.98.
'Frankfurt Airport', O.J. L 72, 11.03.98.
Commission Press Release, 'European Commission ends Frankfurt airport ground handling monopoly', IP/98/27, 14.01.98.
European Commission Press Release: 'The Commission obtains a significant improvement in the supply of ground-handling services at Athens airport: the Commission ends an infringement procedure begun in 1994', IP/97/876, 15.10.97.
'Council Directive 96/67/EC, of 15.10.96, on access to the ground handling market at Community airports', Official Journal, L 272/36, pp.36-45.
European Commission, Directorate General for Transport (DGVII), 'Discussion Document - Revision of Council Regulation 95/93', Brussels, 11.4.96.
'Council Regulation (EEC) No...amending Council Regulation 95/93 on common rules for the allocation of slots at Community airports' (plus Explanatory Memorandum) (1996) -Leaked Document.
European Commission Press Release: Karel Van Miert (Competition Commissioner), Speech to the International Aviation Club, Washington, 8.11.93, IP/93/857.
European Commission Press Release, 'Statement by Mr Van Miert on ground handling services for air transport', Press Release, IP/93/714, 25.8.93.
'Council Regulation (EEC) No 95/93 of 18 January 1993, on common rules for the allocation of slots at Community airports', Official Journal, L 14/1, 12.1.93.
'Third package' of liberalisation - 1992 (EEC) 2407/92 Official Journal No.L 240, 22.8.92 (EEC) 2408/92 Official Journal, L 240, 24.8.92.
Economic and Social Committee, 'Opinion on the proposal for a Council Regulation on common rules for the allocation of slots at Community airports', Official Journal, C 339/41, 31.12.91, pp.41-44.
European Parliament, 'Proposal for a Council Directive on access to the ground handling market at Community airports', with amendments, A3-0215/91,

Official Journal No.C. 267/70, 14.10.91, pp.70-82, and (as above with compromise amendments); Official Journal, C 13/446, 20.1.92, pp.446-456.

European Parliament, Committee on Transport and Tourism, 'Report on the Proposal for a Council regulation on common rules for the allocation of slots at Community airports', A3-0215/91, 31.7.91.

European Commission: 'Proposal for a Council Regulation (EEC) on common rules for the allocation of slots at Community airports', COM(90) 576 final, 30.1.91.

'Second package' of liberalisation - 1990 (EEC) 2344/90 Official Journal 1990, L 217/15.

'Council Regulation (EEC) No.4064/89 of 21 December 1989 on the control of concentrations between undertakings', Official Journal, L 395, 21.12.89.

The *Ahmed Saeed* case, 1989. Case 66/86, *Ahmed Saeed Flugreisen and Silver Line Reiseburo GmbH* v. *Zentrale zur Bekämpfung unlauteren Wettbewerbs* E.V.: [1989] E.C.R. 803 [1990] 4 *Common Market Law Review* 102.

'First package' of liberalisation - 1987. (EEC) 3975/87 Official Journal 1987, L 274/1. 'Block exemptions' granted- (EEC) 3976/87 Official Journal 1987, L 27(EEC) 3976/87 Official Journal 1987, L 274/1.

Nouvelles Frontières case, applying Articles 85 and 86 EC to air transport. Joined Cases 209-213/84, Ministère *Public v. Lucas Asjes:* [1986] E.C.R. 1425, [1986] 3 *Common Market Law Review* 173.

Second Commission Memorandum, European Commission, Memorandum No.2 on Civil Aviation, 'Progress towards the development of a Community air transport policy' COM (84) 72 final.

Challenge to Commission in European Court of Justice - rejected. Case 246/81, *Lord Bethell V. Commission* [1982] E.C.R. 2277, [1982] 3 Common Market Law Review, 300.

European Commission Proposal, applying Articles 85 and 86 EC to the air transport Sector, COM (81) final, 10.9.1981. Official Journal C291/4, as amended by [1982] Official Journal No.C317/3 (rejected by Council).

European Commission: Memorandum on air transport policy, 'Air Transport; A Community Approach', 4.7.79.

'Council Regulation No. 17, 6.2.62. Regulation determining the procedures by which DGIV can execute Articles 85 and 86 of the Treaty'. Official Journal, No.13, 21.2.62, p.204/62.

Documentation prepared for EU officials

Airports Council International (ACI), Europe, 'Response To The European Commission's Discussion Document On The Revision of Council Regulation 95/93, May 1996.

ACI Europe, 'ACI Europe Initial Response Concerning The Potential For Leasing Slots To Airlines', 28.3.96.

ACI Europe, 'Common Rules For The Allocation Of Slots At Community Airports - Review Of And Suggested Revisions to Council Regulation 95/93 of 18 January 1993', 24.11.95.

ACI Europe, 'Comments of ACI Europe on the Draft Council Directive on Access to the Ground Handling Market at Community Airports' (May, 1995).

ACI Europe, 'ACI Europe Resolution on Ground Handling' (adopted by the General Assembly of ACI Europe), Turin, 9.6.94.

ACI Europe, 'Consultations on Ground Handling Services. Response by ACI Europe to the European Commission DGIV and DGVII' (Feb, 94).

ACI Europe, 'ACI Europe Resolution on Ground Handling' (adopted by the General Assembly of ACI Europe), Malta, 10.6.93.

ACI Europe, 'ACI Europe Resolution on Handling' (adopted by the General Assembly of ACI Europe), Madrid, 13.11.92.

ACI Europe, 'ACI Europe Position Paper On Slot Allocation', 19.2.92.

ACI: 'The Economic Benefits of Air Transport' 1992.

ACI Europe, 'ACI Position on Ground Handling at Airports' (and Executive Summary), ACI Europe (Mar, 1991).

Association of European Airlines, Position on Ground Handling, in 'Yearbook' (1995), p.23.

'Comité des Sages for Air Transport to the European Commission, Expanding Horizons', report for the European Commission (DGVII) (Jan, 94).

Coopers and Lybrand, 'The Application and Possible Modification of Council Regulation 95/93 on Common Rules for the Allocation of Slots at Community Airports', Final Report, 17.10.95.

European Regional Airlines Association (ERA), 'Ground Handling in Europe', 22.2.95.

ERA, 'Summary of Key Issues Presented to Right Hon. Neil Kinnock, EU Commissioner for Transport' (April, 1996).

ERA, 'Yearbooks', 95/96; 96/97

Virgin Atlantic, 'A New Approach To Slot Allocation', March 1997.

Virgin Atlantic, 'Briefing Note - Current Air Transport Issues', March 1997.

Intergovernmental documentation, (not directly EEC/EC/EU, in reverse chronological order)

'Agreement Between the Government of the United Kingdom and the Government of the United States of America Relating to Air Services Between Their Respective Territories' (Bermuda Agreement), United Kingdom Command Papers, No.6747, 1946. Also see: 'Agreement Between the Government of the

United Kingdom and the Government of the United States of America Relating to Air Services Between Their Respective Territories'. Verbatim text, published in 1947, in, *The International Law Quarterly,* Vol.1, No.1 (Spring, 1947), pp.286-289.

Chicago Convention, United Kingdom Command Papers, No.6614, 1945. Also see verbatim text in: Appendix A (Part 2), in, Peter Martin and Elizabeth Martin (Shawcross and Beaumont Solicitors) (eds), *Air Law*, Fourth Edition, Vol.2 (London: Butterworths, 1998), pp.A91-A118.

Annual Reports / Company information

ACI Europe, 'Communiqué' (monthly newsletter) (various editions).

BAA, Annual Report (Wilton Road, London), 1996/97.

BAA, 'Take-Off' (staff newsletter) (various editions).

British Airways, Annual Report and Accounts (Speedbird House, Hounslow), 1996/97.

British Airways, 'News' (staff newsletter) (various editions).

Confederation of British Industry, 'Business in Europe' (London: CBI), 1997.

ERA, 'Regional Report' (monthly newsletter) (Chobham, Surrey) (various editions).

London Luton Airport, 'Annual Report and Financial Statements, 1995-1996'.

Manchester Airport, 'Plane Talk' (staff newsletter) (various editions).

Manchester Airport, 'Take-off into the Millennium' (report on second runway project at airport), April, 1997.

Servisair, 'Annual Reports and Accounts' (Manchester),1996.

Conference Proceedings

Bernardine Adkins (Dechert, Price and Rhoads), 'Use of Article 86 to Secure Market Access', paper presented at IBC, 'European Air Transport Services and Airports', conference, Brussels, 6.2.97.

Chris Allen (Head of Competition and Industry Affairs), British Airways, 'The Role of Code Sharing and Franchising in British Airways' Development', paper presented at Royal Aeronautical Society, 'Franchising, Code Sharing and Related Matters', conference, London, 11.2.98.

Chris Allen, 'Comments on slot allocation', paper presented at IBC, 'European Air Transport Services and Airports', conference, Brussels, 6-7.2.97.

Chris Allen, 'Recent developments in slot allocation', paper presented at European Air Law Association, Copenhagen, 8.11.96.

Graham Allen MP (Shadow Minister for Transport), 'Examining the Impact of a Change in UK Government: Presenting Labour's Air Traffic Policy', speech at 'Competitive Strategies for Survival in the Deregulated Aviation Marketplace', conference, London, 24.6.96.

Robert Ayling (Chief Executive), British Airways, 'The Shape of the Airline Industry in the 21st Century', paper presented at the 'FT World Aerospace and Air Transport, Conference', London, 29-30.8.96.

John Balfour (Partner, Frere, Cholmeley, Bischoff, London), 'EC / US airline alliances - who regulates?', paper presented at IBC, 'European Air Transport and Air Services', conference, Brussels, 6.2.97.

John Balfour, 'The EC Ground Handling Directive Explained', paper presented at the Royal Aeronautical Society, 'Slots and Access to Airport Facilities', conference, London, 21.1.97.

Mike Bathgate (Commercial Director, British Regional Airlines), 'Airline Franchising', paper presented at Royal Aeronautical Society, 'Franchising, Code Sharing and Related Matters', conference, London, 11.2.98.

R.G. Brown (EU Economics Studies Group), and F. Sorensen (EU Commission, DGVII, head of air transport section), 'New Developments in European Air Transport', paper presented at the Royal Aeronautical Society, London, 1990.

Frances Butler-Sloss (Partner, Harbottle and Lewis), 'Airline franchising: Some EC law aspects', in P.D. Dagtoglou, J.M. Balfour and S.A. Mirmina (eds), European Air Law Association, Conference Papers, No.9 (London: Kluwer Law International, 1996), pp.45-57.

Chris Castles, Coopers and Lybrand, 'Development of airport slot allocation regulation in the European Community', paper presented at IBC, 'European Air Transport Services and Airports', conference, Brussels, 6-7.2.97.

Philippe Chappatte (Partner, Slaughter and May), 'The Alliance Between SAS and Lufthansa - A Model For Future Co-operation Between Airlines in Europe', paper presented at the European Air Law Association, Copenhagen, 8.11.96.

Mark Clough (Ashurst Morris Crisp, London), 'The Regulation of State Aid in Air Transport', paper presented at IBC, 'European Air Transport Services and Airports, conference, Brussels, 7.2.97.

Bob Cotterill (Head of Economic Policy), UK Civil Aviation Authority, 'The single air transport market - the record to date', paper presented at the European Air Law Association, Madrid, 7.11.97.

Scott Crosby (Partner, Kemmler Rapp Böhlke and Crosby), 'Recent Developments in Slot Allocation', paper presented at the European Air Law Association, Copenhagen, 8.11.96.

Tony Edward (Chief Executive), Messers Dowty Aerospace, 'Commercial Aspects of Aerospace', paper presented at the Royal Aeronautical Society, 'A Rough Guide to Aerospace', conference, London, 18.3.98.

Frank Friedrich (Secretary General), European Chambers of Commerce, 'Practice and legal restrictions on lobbying the European Institutions, conference, European Institute of Public Affairs and Lobbying, European Parliament, Brussels, 23.5.97.

Mark Gerchick (Deputy Assistant Secretary), US Dept of Transportation, 'The US position on air transport liberalisation', paper presented at the 'FT World Aerospace and Air Transport Conference', 29-30.9.96.

Barry Gibson (Group Retail Director), BAA, 'Airport Concepts for the Next Century - The Role of the Airport in the Total Consumer Experience', paper presented at the FT World Aerospace and Air Transport, conference, London, 29-30.8.96.

Malcolm Hart (Managing Director), Gill Airways, 'Partnerships- Our Path to Success' paper presented at Royal Aeronautical Society, 'Franchising, Code Sharing and Related Matters', conference, London, 11.2.98.

Walter Kickert, 'The development in the study of public administration in Europe', paper presented at the Public Administration Committee, Chartered Institute of Public Finance and Accounting, London, 17.1.97.

Dinos Kyrou, 'Route support- a paper for Airport Strategy and Marketing', presented at the ASM Airport Routes conference, London, 21-22.9.98.

Yves Morier (Regulations Director), Joint Aviation Authorities, 'The Joint Aviation Authorities (JAA) Safety Strategy Initiative', paper presented at IBC, 'Aviation Safety Management', conference, London, 15.5.98.

Michael Niejahr (Administrator, DGVII), 'Recent developments in slot allocation', paper presented at the European Air Law Association, Copenhagen, 8.11.96.

Hugh O'Donovan (Partner, Wilde Sapte, UK), 'Dealing and Trading in Slots', paper presented at the Royal Aeronautical Society, 'Slots and Access to Airport Facilities', conference, London, 21.1.97.

Trevor Soames (Partner, Norton Rose),'Opportunities and Challenges in the Aviation's New World Order: The Application of EU Competition Rules to Air Routes from the EU to Third Countries', paper presented at the European Air Law Association, Vienna, Nov. 1998.

Trevor Soames, 'Access to airport facilities, the impact of competition law', paper presented at the Royal Aeronautical Society, 'Slots and Access to Airport Facilities', conference, London, 21.1.97.

Trevor Soames, 'Ground Handling Liberalisation', paper presented at the IBC, 'European Air Transport Services and Airports', conference, 6-7 2.97.

David Starkie (advisor to DGVII on slot allocation), 'Recent Developments in Slot Allocation: What Can We Learn From The US Market In Airport Slots?', paper presented at the European Air Law Association, Copenhagen, 8.11.96.

David Starkie, 'Allocating Airport Slots: A Role for the Market?', paper presented at the Institute of Economic Affairs, London, 12.2.97.

Le Thi Mai (Head of Infrastructure and Environment), AEA, 'Ground handling liberalisation', paper presented at the IBC 'European Air Transport Services and Airports', conference, Brussels, 6-7.2.97.

Karel Van Miert, 'Competition policy in the air transport sector', speech at the Royal Aeronautical Society, London, 9.3.98.

Kees Veenstra, 'Discrimination Under EC Slot Allocation Rules In Favour Of New Entrants At Congested Hub Airports', paper presented at the Royal Aeronautical Society, 'Slots and Access to Airport Facilities', conference, London, 21.1.97.

Robert Webb QC (Head of Chambers, 5 Bell Yard, London), 'The legal implications of slot trading', paper presented at the Royal Aeronautical Society, 'Slots and Access to Airport Facilities', conference, London, 21.1.97.

Hugh Welburn (Economic Adviser to Virgin Atlantic), 'Dealing in slots, paper presented to the Institute of Economic Affairs, London, 12.2.97.

Stephen Wilks, 'Models of European Administration', Conference of the European Group of Public Administration, 2-5.9.92.

Other Studies

House of Commons, Transport Committee, Sixth Report, 'The Proposed Alliance between British Airways and American Airlines', 30.7.96.

IATA, Scheduling Procedures Guide (Geneva: IATA, 1994).

Dinos Kyrou, 'Memorandum on Slot Allocation at Community Airports', University of Aberdeen (Memorandum for Keith Bradley, Shadow Transport Minister), 13.12.96.

Ian Stockman, 'A Comparative Study of User Costs at Selected European Airports' (Air Transport Research Group), Cranfield University, 1996.

UK Dept. of Transport, Aviation Group, 'EC Aviation Liberalisation Information', Summer 1996.

UK Dept. of Transport, 'US Airline Deregulation: The Lessons for the European Community', Working Paper No.109, September 1989.

Web Sites

ACI's main web site is at: www.airports.org

ACI Europe's web site is at: www.aci-europe.org

The AEA has no web site.

American Airlines' web site, (which contains substantial legislative information): www.americanair.com

British Airways' web site: www.british-airways.com

DGIV's web site can be found on:

www.europa.eu.int/en/comm/dg04/dg4home.htm
DGVII's web site: www.europa.eu.int/en/comm/dg07/index.htm
Easyjet's web site contains a great deal of legal information concerning the alleged anti-competitive pricing policy of British Airways. It can be found on: www.easyjet.com
ERA's web site can be found at: www.eraa.org
European Integration online Papers, (EIoP): www.eiop.or.at/eiop
European Union: Official web site: www.europa.eu.int
IATA's web site: www.iata.org/index.asp
Virgin Atlantic's web site: www.fly.virgin.com

Newspapers

The Daily Telegraph
The European
The Financial Times
The Guardian
The Independent
The Independent On Sunday
The Observer

For Product Safety Concerns and Information please contact our EU representative GPSR@taylorandfrancis.com Taylor & Francis Verlag GmbH, Kaufingerstraße 24, 80331 München, Germany

Batch number: 10725909

Printed by Printforce, the Netherlands